Practical Statistics
for
Experimental Biologists

Practical Statistics for Experimental Biologists

A. C. WARDLAW

Professor of Microbiology
University of Glasgow

A Wiley–Interscience Publication

JOHN WILEY & SONS

Chichester · New York · Brisbane · Toronto · Singapore

Library of Congress Cataloging in Publication Data:
Wardlaw, A. C.
  Practical statistics for experimental biologists.
  'A Wiley–Interscience publication.'
  Bibliography: p.
  Includes index.
  1. Biometry.   2. Biology, Experimental — Statistical
methods.   I. Title.
QH323.5.W37   1985   574'.0724   85-3147
ISBN 0 471 90737 5
ISBN 0 471 90738 3 (pbk.)

British Library Cataloguing in Publication Data:
Wardlaw, A. C.
  Practical statistics for experimental biologists.
  1. Biometry
  I. Title
  519.5'024574   QH323.5
  ISBN 0 471 90737 5
  ISBN 0 471 90738 3 pbk

Phototypeset by Dobbie Typesetting Service, Plymouth, Devon
Printed and bound in Great Britain

Contents

Preface

Why another book on biostatistics? Well, that question could be bounced back by asking why there are so many biostatistics texts already in existence, considering that several of them are widely acclaimed as standard works. One major difference between this book and most of the others is that the author is not a qualified statistician, but a laboratory scientist with a background in organic chemistry, and subsequent academic and industrial experience in microbiology and immunology, with excursions into pharmacology, genetics and ecology. So what follows in these pages is the author's own personal distillate of statistical insights and techniques which he, as a laboratory biologist, believes may be useful to others with similar interests.

The book sets out to be 'user-friendly', to take a vogue expression which happens to be appropriate. For example, in considering the statistical procedures which may be used with count data — like bacterial colony counts, blood cell counts and radioactivity counts — the chapter is headed *How to deal with count data*. Whereas a 'real' statistician might have called it *Applications of the Poisson Distribution*. This approach is used throughout. The reader is assumed to be a busy, and probably impatient, experimental scientist in biology, biochemistry, physiology or pharmacology who is generating data in the laboratory and who wants to 'do the correct thing', statistically speaking, but without having to undergo a long apprenticeship in statistical theory. So the book is deliberately light on theory and is not for the reader who wants to see equations being derived from first principles. To take an automobile analogy, this book aims to be similar to a do-it-yourself manual on how to drive and maintain a motor car, but without first having to understand the thermodynamics of hydrocarbon combustion.

Throughout the book there is repetition of an essential message which statisticians regard as so self-evident as not to need emphasis, but which experimental biologists often tend to ignore. It is the exhortation that:

The statistical procedures have to be built in BEFORE the experiment is done, and should not simply be regarded as the number-crunching that is carried out after the data have already been gathered.

The benefit for those readers who take this advice is that even if they themselves cannot do the statistical calculations from their own experiments, they should at least have acquired acceptable data for processing by a professional statistician.

I am indebted to many individuals and sources for the material in this book. The standard texts of Bliss (1967), Campbell (1974), Colquhoun (1971), Finney (1971) and Snedecor and Cochran (1967) have been invaluable reference works. Among my statistician colleagues, I am particularly appreciative of discussions with D. B. W. 'Bill' Reid and Gerald Van Belle, formerly of the University of Toronto, and with David McLaren of Glasgow University. However, no blame should go to any of them for my shortcomings and errors. I also thank Dr Roger Parton and Miss Fiona Sidey of Glasgow University for reading the manuscript from the standpoint of a consumer. The book would not have been possible without permission to copy having been granted by the numerous sources which are acknowledged at the appropriate places in the text. I am grateful to the Literary Executor of the late Sir Ronald A. Fisher, F.R.S., to Dr Frank Yates, F.R.S. and to Longman Group Ltd, London, for permission to reprint Tables III, IV, VII, IX and IX2 from their book *Statistical Tables for Biological, Agricultural and Medical Research* (6th Edition, 1974). The verse from John Masefield's poem 'Truth' is reprinted with permission of the Society of Authors, as the literary representatives of the Estate of John Masefield, and with permission of Macmillan Publishing Company from *Poems* by John Masefield. Copyright 1912 by Macmillan Publishing Company, renewed 1940 by John Masefield. Finally, I thank the publishers for their patience and Mrs Anne Mosson and Mrs Margaret White for producing the typescript and for enduring numerous revisions.

A.C.W.
Glasgow, 1984

Man with his burning soul
Has but an hour of breath
To build a ship of Truth
In which his soul may sail—
Sail on the sea of death
For death takes toll,
Of beauty, courage, youth,
Of all but Truth.

John Masefield

1
A Simple Laboratory Exercise

The notion that experiments and other research investigations can be conducted statistically or non-statistically, at the will of the investigator, is firmly held by many: it is usually entirely false.

D. J. Finney

1.1   Coming to Grips with Statistics

It is difficult to be an experimental biologist and not use a pipette or similar device to deliver measured quantities of test fluids. Mostly we take the accuracy of these dispensing instruments completely for granted but, from time to time, it may be highly salutary actually to check this assumption. So what follows is a simple laboratory exercise which should prove most illuminating, first as a check on the accuracy of your pipettes but, more importantly, as an introduction to statistical methodology. It is an exercise that every biology student and laboratory scientist should endure at least once, and I think you may be surprised at how many different statistical techniques can be illustrated by it. However, you should appreciate that the data obtained from this exercise are to be regarded imaginatively as a simple example of the many different types of quantitative and repetitive measurements that are generated in a laboratory on all kinds of different instruments and from many different experimental systems. So the statistical methods we develop in this example should be seen as having very wide and general applications. To base the exercise on pipettes is just to make it as simple as possible from a technical standpoint.

Those who are more deeply interested in an experimental investigation of the precision and accuracy of using the Pasteur pipette, the Eppendorf pipette and Microtiter diluters for dispensing fluids should consult Kolk-Vegter *et al.* (1973) for particulars.

1.2   Standardizing a Pipette by Weighing the Deliveries

Although the pipette is a device for measuring volume, an easy way to check its accuracy is to weigh the delivery of a liquid of known specific gravity. An ordinary analytical balance can weigh to 0.1 mg, which is 1/10,000 of the contents of a 1 cm^3-pipette filled with water. This level of accuracy is much higher than one could hope to achieve in ordinary usage of the pipette, where the smallest calibration marks are at 0.01-cm^3 (i.e. 10-mg) intervals. Note that, for biological purposes, we can take the cubic centimetre (cm^3) and the millilitre (ml) as being the same (1 ml = 1.000028 cm^3).

I now want you, the reader, actually to do or to imagine yourself doing the following simple exercise: take 3 ordinary 1-cm^3 graduated pipettes and make, as carefully as possible, 3 deliveries of 1 cm^3 of air-free distilled water at 20°C from each. The deliveries are to be made into a weighing bottle which is weighed empty, to obtain the tare weight (unless you have an automatic tare balance), and then reweighed after each delivery—a total of 10 weighings. The whole job should take no more than about 20 minutes. To conform to safety requirements, you will no doubt use some kind of attachment on the pipette so as to avoid mouth-pipetting. Any lack of dexterity in using this device will be revealed.

1.3   Purpose, and therefore Design, of the Exercise

There are 5 initial questions to be answered:
   (1) What is the *average* weight of water in the 9 deliveries?
   (2) How much *scatter* or *variation* is there in the 9 deliveries?
   (3) How accurate are the 9 deliveries? This is not the same as Question 2.
   (4) Are there *significant differences* in the mean volumes delivered by the 3 pipettes?
   (5) How should the benchwork be done so as to avoid confusing any volumetric differences between individual pipettes with a possible training or

fatigue effect of the operator, e.g. using the 3rd pipette more (or less) carefully than the 1st?

These are questions of a type that arise frequently in biological experiments. Therefore the experience we gain in this simple test may well find immediate application in your own current laboratory work. I hope that at least some of it will!

How then should the weight readings be taken? There are several points to discuss before actually doing the laboratory work.

Method 1 (how many readers would probably proceed if left to their own devices)

Take the tare weight of the weighing bottle; then make 3 deliveries and weighings from pipette A, followed by 3 from pipette B, finishing up with 3 from pipette C. This will certainly give a nice tidy array of results, AAABBBCCC, but is this necessarily the best method? You could make the obvious criticism that the exercise, if done this way, is not suited to answer Question 4, namely whether any apparent differences between the pipettes themselves are *real*, or whether they might have arisen from the experimenter varying in care or expertise while progressing through the 9 measurements. It would be like taking a person into a shooting alley to fire 3 shots from each of 3 rifles. If the shots were closer to the bull's-eye with the third rifle than with the first, how would you know whether this was due to the last rifle being more accurate, or because the shooting improved progressively with successive shots?

Thus, with both the pipettes and the rifles we have potentially two *independent* sources of variation (this is an important and basic statistical idea):

(a) variation due to intrinsic physical differences between individual pipettes or rifles, and

(b) variation originating in the wobbly hand or eye of the human operator.

If we wish to be able to detect and assess the importance of these two quite separate sources of variation, then the experiment must be done in such a way that the two sources are not confused or *confounded* (to use the statistician's term). In other words the two influences — if we think they may affect the results — should not be mixed up in such a way that we are unable to separate them afterwards.

Method 2 (more sophisticated)

This takes account of the above criticisms and is the method we shall use. The essential feature is that the 9 deliveries are to be made in a *predetermined random sequence*, such as AACBCABCB. We shall discuss below how randomness is achieved, because it is not just a matter of arranging the letters to look random. We could also introduce two additional worthwhile practical refinements.

(1) To have the pipettes unmarked, but individually identified by being placed on labelled sheets of paper, kept out of sight of the pipetting operator and passed

from behind by an assistant. The assistant would choose a particular predetermined random sequence which would not be disclosed to the operator until the exercise had been completed. Thus the operator would not know whether any particular delivery was being made with pipette A, B or C.

(1) Initially to wet each pipette internally so that the operator would not know whether a particular pipette had been used previously.

In adopting this *completely random* layout for the exercise we are not excluding other possibilities and, for example, Chapter 7 will discuss two other designs, namely the Randomized Block and the Latin Square, for each of which an argued case could be made.

At this stage many readers may well feel that an unnecessarily elaborate ritual is being set up for a very simple exercise. But the intention is to emphasize the crucial point that:

If you intend to use statistical methods to analyse the *results* of an experiment, then these methods must be built in *before* the experiment is done.

Otherwise your work will be on the same level of inadequacy and incompetence as, say, of using unsterilized culture media for growth of bacterial cultures. It is an unfortunate fact that many experimental biologists blithely apply statistical methods to the analysis of data which have been gathered without thought of possible sources of bias and without any attempt at randomization. Sad, isn't it? And it should be no surprise therefore that statistics, as an intellectual discipline, tends to get a bad reputation: 'You can prove anything with statistics'; 'There are three kinds of lies—lies, damned lies and statistics', etc. Any tool or device used incorrectly can give bizarre results and statistical methods are no exception.

So without further ado, let us accept that there are good reasons why our 9 pipette deliveries should be collected in a *predetermined random sequence*, while keeping in mind that this is only one of a number of acceptable options. As we shall see repeatedly in this book, the experimental scientist has to develop the knack of treading the middle ground between achieving 'full statistical purity'—which, like other forms of virtue, is difficult and troublesome to attain—and doing poorly designed and unanalysable experiments. This middle way requires knowledge, judgement and experience. And do not think that the issue can be dodged by saying 'A well designed experiment does not need statistics'. If the experiment is *indeed* well designed, it will necessarily include statistical principles such as replication and avoidance of bias.

1.4 Arranging a Random Sequence

Statisticians are very fussy about randomness—to a degree which tends to leave biologists rather bemused (although not necessarily *amused*!). For example, you should not attempt randomization simply by putting down what looks like a random sequence of A's, B's and C's. Nor would it be fully acceptable to arrange our 9 random deliveries by drawing cards from a shuffled pack of 9 numbered or lettered cards. In both of these cases we would be liable to have our efforts scorned as 'quasi-random devices'. It is best not to argue about this but just accept that to achieve 'Statistical Purity' in the matter of randomness, you should use a *Table of Random Numbers*. An example of such a table is given in Appendix A1, part of which is reproduced in Table 1.1 for convenience. It is

Table 1.1   Excerpt of the first two lines of the table of random digits given in Appendix A1

20 17	42 28	23 17	59 66	38 61	02 10	86 10	51 55	92 52	44 25
74 49	04 49	03 04	10 33	53 70	11 54	48 63	94 60	94 49	57 38

actually a table of random *digits* from 0 to 9, and what is meant by 'true randomness' is that at any geographical point in the table there is an equal probability of finding any of the 10 digits from 0 to 9. The table may be read horizontally, vertically, or any other way and the digits can be taken singly, in pairs, triplets, etc. depending on how many numbers or objects are to be randomized. The way the table is printed in groups of 4 digits in vertical columns has no purpose other than to make it easier to read. Thus, the random digits in the first horizontal line are to be read as 2, 0, 1, 7, 4, 2, 2, 8, 2, 3, etc. although they are printed as 2017, 4228, 2317, etc. to make scanning them easier. The table can be entered at any point and, if you are making routine use of it, you would normally start to extract each new set of random numbers beyond where the previous set ended.

In our case, let us assign digits 1, 2 and 3 to pipette A; 4, 5 and 6 to pipette B; and 7, 8 and 9 to pipette C. If we now take the first horizontal line of the table, we get the sequence and code:

2	0	1	7	4	2	2	8	2	3	1	7	5	9	6
A	–	A	C	B	–	–	C	–	A	–	–	B	C	B

where we put a blank for each digit that either has already been found or, as with zero, is not wanted. We shall therefore take deliveries from the pipettes in the sequence AACBCABCB. This we can accept as a true random sequence and obviously is one of a large number of such sequences.

Note that by chance we could finish up, from the random number table, with a regular sequence such as AAABBBCCC. However, it is legitimate to discard such sequences and take another set provided we have declared beforehand our intention of so doing.

There are various other ways of extracting numbers from a table of random numbers. For example, if several score of patients had to be assigned at random to two treatment groups in the clinical trial of two drugs, A and B, we could link Group A with odd digits and Group B with the even digits and zero. Thus as successive patients arrived for treatment, they would be assigned (reading horizontally along the first line of the table) to groups BBAABBBBBAAAAABB, etc. At the end of row 1, there would be 18 in

A and 22 in B and they would have been allocated quite at random. As the total number of patients grew, the numbers accumulating in A and B should be very closely similar '*within the limits of random-sampling fluctuations*'.

1.5 Tabulating Results

Two of the commonest types of mistake in statistical calculations are *simple arithmetical errors* and *errors in copying numbers*, especially the very large numbers which tend to arise at intermediate stages in statistical calculations. It is therefore important to cultivate good work habits which keep these to a minimum. One such habit is the tabulation of data in a properly constructed table, preferably on ruled or large-squared paper. To be safe therefore in our pipetting experiment, we need first a blank table for recording the weights of water delivered in the randomized sequence in which they are actually obtained. We then need a second table to sort out and arrange systematically the three readings from each pipette. If we try to telescope the two tables together we are liable to get into a mess.

Table 1.2   The collection and initial processing of the results of the pipetting exercise

| | | Weight of water delivered | | |
Pipette (1)	Balance reading (g) (2)	in grams (3)	in milligrams (4)	in 'working milligrams' (5)
B	34.1198	0.9820	982.0	12.0
C	33.1378	0.9759	975.9	5.9
B	32.1619	0.9947	994.7	24.7
A	31.1672	0.9927	992.7	22.7
C	30.1745	0.9977	997.7	27.7
B	29.1768	0.9816	981.6	11.6
C	28.1952	0.9948	994.8	24.8
A	27.2004	0.9868	986.8	16.8
A	26.2136	1.0075	1007.5	37.5
Tare	25.2061	0.0000		

Column 1 records from the bottom upwards the predetermined random sequence of pipette usage.
Column 2 records the successive balance readings, also from the bottom upwards (to facilitate subtraction), after each delivery.
Columns 3, 4, 5 give the weights of water delivered in grams, milligrams and 'working milligrams', the last being obtained by subtracting the constant 970.0 from each value in column 4.

Table 1.2 is an example of a table in which the raw data are gathered. Note that in this case the weights are accumulated from the bottom upwards, so as to facilitate subtraction of the successive values in column 2. Column 1 is the predetermined random sequence in which the pipettes were used. Column 2 then records the 6-figure balance readings, starting at the bottom with the tare weight

'Look — if you have five pocket calculators and I take two away, how many have you got left?'
(Reproduced through the courtesy of *Punch*)

of the empty vessel. In the next section we shall start to process the figures, after some brief advice about choosing a calculator.

1.6 Choosing a Calculator

The calculations in this book assume that the reader possesses only a simple pocket calculator with the following features: one or more memories; square and square root; ability to transform numbers to logarithms to the base 10 and vice versa, and the ability to generate factorials; a programme for calculating the mean and standard deviation and, most important, a facility for displaying ΣX and ΣX^2.

No attempt is made in this book to introduce computer programs or to make use of statistics software such as is widely available for the numerous types of microcomputer now on the market. To venture into either of these areas would require a book in itself and would go far beyond what is possible within these pages. However, it is assumed that the reader who is going to do statistical calculations routinely will get access either to a mainframe or a microcomputer and will acquire or develop a portfolio of the necessary programs. Some useful references for this purpose are: Cooke *et al.* (1982), Fegan and Brosche (1979), Lee and Lee (1982) and Sokal and Rohlf (1969).

1.7   Arithmetic Mean and Standard Deviation

1.7.1   Preliminary simplifications

When confronted with a set of data like the 4- and 5-figure results in column 3 of Table 1.2, the first step to consider is *simplification*. But this does *not* mean rounding off some of the terminal digits, because we would then be discarding needlessly some of the inherent accuracy of the results, which would negate the purpose of the exercise. Instead, there is another approach which allows us to do the statistical calculations with simpler numbers but without downgrading the accuracy of the original data. This is done in three discrete steps as shown in columns 3, 4 and 5 of Table 1.2. In column 3 we tabulate the weight readings in grams to 4 decimal places and, indeed, we might wish to work directly with these figures. However, the results are easier on the human eye if transformed from grams to milligrams as in column 4. Again note that we have not rounded-off terminal digits as this would cause unwanted loss of information.

In column 5 of Table 1.2 we use another simple device, namely subtracting a constant. The nice round number of 970 has been chosen for subtraction from each of the milligram values in column 4, so that we are left with a set of small positive numbers from 5.9 to 37.5, which I have called 'working milligrams' and on which we shall do our statistical calculations. At the end of the calculations we can then add back the constant 970 to bring everything back to 'real' milligrams.

Let us then re-emphasize that in deriving the 'working milligrams' we have not thrown away any of the information such as would occur if we had only recorded the weighings to 3 instead of 4 decimal places. The simplifications we have used are purely arithmetical devices which in no way downgrade the accuracy of the raw data.

It cannot be repeated too often that even professional statisticians make arithmetical and copying errors. So it is important to arrange the work with proper tabulation followed by an orderly sequence of steps and neat handwriting so as to eliminate these intellectually trivial but potentially ruinous errors. You should also be prepared to check *all* calculations by repeating them, and by having built-in cross-checks at crucial points where possible. This advice should be repeated and repeated. You must check and double-check everything.

Table 1.3   'Working milligram' weights arranged according to pipette, together with the means

	Pipette A	Pipette B	Pipette C
	37.5	11.6	24.8
	16.8	24.7	27.7
	22.7	12.0	5.9
Mean value in working mg	25.67	16.1	19.47

The final step in the simplification and tabulation of the results is to unshuffle the randomization of the 'working milligram' readings and arrange them according to which pipette they came from, as in Table 1.3. Initially, however, we shall treat the data as a set of 9 replicate measurements and temporarily ignore that they arose from 3 separate pipettes.

1.7.2 Arithmetic mean

The arithmetic mean, or simply the 'mean', is the same as the 'average'. We use the qualifying term arith*metic*, with emphasis on the third syllable, to distinguish it from the *geometric* mean (Section 2.4.5).

At this point it is necessary to introduce some symbols. Our 9 individual items of data (y) — the 'working' figures in column 5 of Table 1.2 — will be designated y_1, y_2, . . . , y_9, and the number of these items of data by N. The mean ($\bar{y}$), pronounced 'y-bar', is then given by

$$\bar{y} = \frac{\Sigma y}{N} \text{ which may also be written } \Sigma y/N \quad \ldots\ldots\text{Eq. 1.1}$$

where capital Σ, 'sigma', is the summation sign and is simply a command to add up all the 9 y-values, i.e.

$$\Sigma y = (37.5 + 16.8 + 24.8 + \ldots + 12.0) = 183.7$$
$$\Sigma y/N = 183.7 \div 9 = 20.4111,$$

which we can round off to 20.41.

I suggest that you work through these calculations as they occur in the text to gain practice and familiarity and as a base for the more complicated equations that come later and which will be given with less intermediate working and fewer explanations.

If we now add 970 to 20.41 it gives the mean of the 9 pipette deliveries in 'real' milligrams as 990.41 mg. Note that the calculator may give us the result to many more decimal places than are worth recording and we should avoid a spurious appearance of accuracy by excessive zeal in copying lengthy decimals. Here, two digits after the decimal point are sufficient because they give us the mean with one more digit than each individual reading on which the mean is based. This seems reasonable since the mean is the average of 9 attempts and should therefore be a 'better' value than any single attempt at pipetting.

1.7.3 Standard deviation by the 'stone-age' method

When confronted with a series of replicate values such as the 9 pipette deliveries in Table 1.2, we can ask various questions, but two obvious ones are: What is the mean? and How much *variation*, *spread* or *scatter* is there?

There are several ways to assess variation or variability in replicate observations, but by far the most useful is the quantity known as the *standard deviation*, which is given the symbol s. The formula for calculating the standard deviation is:

$$s = \sqrt{\frac{\Sigma(y - \bar{y})^2}{(N-1)}} \quad \dots\dots\dots\dots\dots\dots\dots \text{Eq. 1.2}$$

To show how the formula works, we shall calculate the standard deviation of the 9 'working milligram' results in Table 1.2 by direct substitution. Then we shall show how the calculation can be streamlined (Section 1.7.4) and re-streamlined (Section 1.7.5) so that the whole operation becomes quite painless and routine. Meanwhile let us proceed with what I have called the 'stone-age' method of calculating standard deviation, as an exercise in simple arithmetic.

Table 1.4 Calculation of the standard deviation of the 9 pipette deliveries from Table 1.2 using the formula

$$s = \sqrt{\frac{\Sigma(y - \bar{y})^2}{(N-1)}}$$

Weight in 'working' mg (y)	Deviation from overall mean $\bar{y} = 20.41$ ($y - \bar{y}$)	Deviation squared ($y - \bar{y})^2$
12.0	-8.41	70.7281
5.9	-14.51	210.5401
24.7	4.29	18.4041
22.7	2.29	5.2441
27.7	7.29	53.1441
11.6	-8.81	77.6161
24.8	4.39	19.2721
16.8	-3.61	13.0321
37.5	17.09	292.0681
	Total $= 0.01$	$\Sigma(y - \bar{y})^2 = 760.0489$

$$\frac{\Sigma(y - \bar{y})^2}{(N-1)} = 760.0489/8 = 95.0061$$
$$s = \sqrt{95.0061} = 9.7471$$

First we shall transfer column 5 of Table 1.2 into the left column of Table 1.4 to provide the starting material. As shown in Table 1.4, each value (y) is subtracted from the mean ($\bar{y}$), which is 20.41, to give the central column of figures ($y - \bar{y}$). These are the *individual deviations from the mean*, with plus and minus signs. The sum of the pluses and minuses should be zero, but if it is not, there may be simple arithmetical or copying errors or excessive rounding off of decimals in recording the mean. Here it is close enough to zero to be satisfactory. The third column shows $(y - \bar{y})^2$, *the squares of the deviations.* Note that the minus signs disappear on squaring; and by adding up the final column we get $\Sigma(y - \bar{y})^2$, the *sum of the square of the deviations* equal to 760.0489.

To complete the calculation, we divide by $(N-1)$ which is 8, to get 95.0061, whose square root is 9.7471, the standard deviation. This value of the standard deviation has the units of 'real' milligrams despite the calculation being done in 'working milligrams'. This is because the numerical value of the standard deviation is not affected by subtracting a constant, although the value of the mean *is* affected, i.e. standard deviation is a measure of the *scatter* of values around the mean, and is not influenced by subtracting the same constant from each observation in the group.

There are several questions you may want to ask about standard deviation, such as:

(1) Why does the formula use the *squares* of the deviations, and not just the deviations themselves?

(2) Why is the divisor $N-1$ and not N, considering that there are N observations?

(3) In what sense is the deviation 'standard'?

(4) Is there any mental picture that we can form for the extent of variation implied by the numerical value of the standard deviation?

The answers to the first 3 questions require quite deep delving into statistical theory and are therefore outside the scope of this book. Question 4 is dealt with in the next chapter.

1.7.4. A better formula for standard deviation

Although Eq. 1.2 is the simplest way to *write* the formula for standard deviation, it is not the most convenient to use in the actual calculation. In fact, let's be frank: that last calculation we did was rather laborious. A better formula emerges by rewriting the top line under the square root, for it can be shown algebraically that $\Sigma(y - \bar{y})^2$ is mathematically identical to $\Sigma y^2 - (\Sigma y)^2/N$. This gives the improved version of the same formula as:

$$s = \sqrt{\frac{\Sigma y^2 - (\Sigma y)^2/N}{(N-1)}} \quad \dots\dots\dots\dots\dots\text{Eq. 1.3}$$

Table 1.5 presents the calculations using this better formula. You will note that we no longer have to calculate the mean and subtract each individual result from it. This is a big help because your calculator should be able to accumulate Σy and Σy^2 simultaneously in two of its memories as each number is entered and squared.

You should therefore work through the calculation in Table 1.5 to become familiar with the method and to satisfy yourself that it gives the same value

Table 1.5 Calculation of the standard deviation of the pipette deliveries from Table 1.2 using the formula

$$s = \sqrt{\frac{\Sigma y^2 - (\Sigma y)^2/N}{(N-1)}}$$

Weight in 'working' mg (y)	y^2
12.0	144.00
5.9	34.81
24.7	610.09
22.7	515.29
27.7	767.29
11.6	134.56
24.8	615.04
16.8	282.24
37.5	1406.25
$\Sigma y = 183.7$	$\Sigma y^2 = 4509.57$

$(\Sigma y)^2 = 183.7^2 = 33745.69$
$(\Sigma y)^2/N = 33745.69/9 = 3749.5211$
$\Sigma y^2 - (\Sigma y)^2/N = 4509.57 - 3749.5211 = 760.0489$
Dividing by $(N-1)$ gives 95.0061
Take square root, gives $9.7471 = s$

Note: the term $(\Sigma y)^2/N$ is often referred to as the *correction factor*.

Table 1.6 Calculation of the standard deviation of the pipette deliveries using untransformed gram data and Procedure 2

y	y^2
0.9820	0.96432400
0.9759	0.95238081
0.9947	0.98942809
0.9927	0.98545329
0.9977	0.99540529
0.9816	0.96353856
0.9948	0.98962704
0.9868	0.97377424
1.0075	1.01505625
$\Sigma y = 8.9137$	$\Sigma y^2 = 8.82898757$

$(\Sigma y)^2/N = (8.9137)^2/9 = 8.828227521$
$\Sigma y^2 - (\Sigma y)^2/N = 8.82898757 - 8.828227521$
$= 0.0007600489$
Dividing by $(N-1)$ gives 0.0000950061
Take the square root to get $s = 0.0097471$ g
$= 9.7471$ mg

of *s* as before and with considerably less effort and possibility of mistakes.

Before leaving this section, let us go back and consider the virtue of doing our calculations of *s* in 'working milligrams' as we have done, rather than in the original grams in which the data were first gathered. If we had not bothered to make the simplifications by converting to milligrams and subtracting 970, the calculation of *s* by the improved formula would be as in Table 1.6. Note that we get the same answers, $s = 9.7471$ mg, at the end, but there are some horribly long numbers to deal with in the intermediate stages and it is very easy to make copying errors with large numbers such as these. So although the transformation to 'working milligrams' involved extra labour at the beginning, it was probably worthwhile as a simplification during the subsequent work — certainly for initial instructional purposes, as here.

1.7.5 Routine calculation of mean and standard deviation

Your calculator should have a built-in program for calculating mean and standard deviation, and all that is required is to enter the 'working milligrams' and press the appropriate button after each entry. This you should now do with the data of column 1 of Table 1.5 and satisfy yourself that the same answer is obtained by this method as by use of the more tedious procedures we have just employed. So, although the calculator program will henceforth be our normal method of obtaining the mean and standard deviation, it is worthwhile at least once, for instructional purposes, to work through the more tedious and error-prone methods of Sections 1.7.2–1.7.4. One final word of caution: your calculator is probably programmed to calculate standard deviation with either N as the divisor under the square root or $(N-1)$. Unless otherwise advised, you should always use the $(N-1)$ option.

1.8 Precision and Accuracy

These two terms are often used loosely and incorrectly, although they have distinct and different meanings. *Accuracy* contains the idea of agreeing with an accepted reference measurement or 'target' value, whereas *precision* is concerned with closeness of reproducibility in replicate observations or tests. Fig. 1. 1 uses target-shooting to illustrate these points. In Target A, where all the shots are tightly grouped within the bull's-eye, both precision and accuracy are of a high order. In Target B, the precision is just as high because the shots are in an equally tight group, but the *accuracy* is lower. This could result from some consistent error in the rifle barrel or the sights, or because of a steady cross-wind or some consistent fault in aiming. Any of these factors could affect *accuracy* without necessarily reducing precision. The corresponding situation in our pipetting exercise might arise if we used a pipette which had been manufactured with an inherent error in calibration. It might also develop from a consistent fault by the operator, e.g. using a 'delivery' pipette as if it were of the 'blow-out' variety, or adjusting the meniscus of the liquid in the pipette

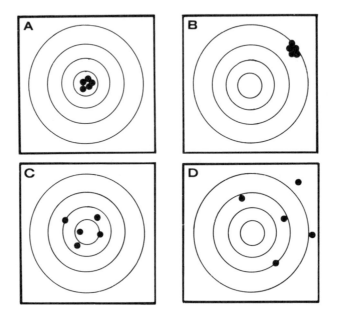

Fig. 1.1 Precision and accuracy illustrated by target-shooting: a, precision and accuracy both very good; b, precision as high as in a, but accuracy poor; c, quite accurate, but precision lower than a or b; d, both precision and accuracy are poor

to be consistently at the wrong level in relation to the graduation marks.

Target C is reasonably *accurate* (more so than B), in that the shots are equally placed around the bull's-eye, but the *precision* is not as good as in A or B because there is more scatter about the mean. Such a result would be expected from an accurate rifle being fired by someone with shaky aim. In pipetting, the analogous result would be given by an accurately calibrated pipette used with inadequate consistency. In target D both precision and accuracy are poor as shown by the wide scatter of shots well away from the centre.

Returning again to the pipetting, the standard deviation is our measure of *precision*, or extent of scatter about the mean value. A small standard deviation corresponds to a 'tight group' and a large standard deviation to widely scattered shots. On the other hand, the *accuracy* of our pipetting is indicated by how close our individual shots and our mean value of 990.41 mg are to the 'target', or accepted, value of 998.21 mg for the weight of 1 cm^3 of pure, air-free water at 20°C and 760 mm atmospheric pressure. The difference between the centre point of our 'shooting' and the 'bull's-eye' is 7.80 mg. How do we decide whether this is 'accurate' or not?

Basically, it is a question of how we define 'accurate'. The first point to make is that accuracy is a matter of degree rather than an absolute quantity. With target-shooting, an 'accurate' shot is one that hits the bull's-eye: it does not have to pass through the geometrical dead centre in order to be considered 'accurate'. Similarly with pipetting, the physical equivalent of a bull's-eye of particular diameter is the *tolerance limit of volume* within which we would be prepared to consider the pipetting as worthy of the term 'accurate'. In regard to the pipettes themselves, the British Standards Institution (1962) prescribes two classes of accuracy for graduated 1 ml pipettes. For Class A pipettes the tolerance, or maximum allowed error of calibration, is ± 0.006 ml, which is equivalent to 6.0 mg of water. For Class B it is ± 0.01 ml equivalent to 10 mg of water. This would give tolerance ranges of:

$$998.21 \pm 6.0 = 992.21 \text{ to } 1004.21 \text{ for Class A accuracy}$$
$$998.21 \pm 10.0 = 988.21 \text{ to } 1008.21 \text{ for Class B accuracy}$$

We can now consider the accuracy of our pipetting in terms of
(a) accuracy of the individual deliveries; or
(b) accuracy of the average of our 9 deliveries; or
(c) accuracy of the mean of the 3 deliveries from each pipette.

In regard to the individual deliveries, reference to column 4 of Table 1.2 shows that only 4 of the 9 deliveries achieved Class A accuracy and a further 1 achieved Class B. In regard to our overall average value of 990.41, this falls within the limits of Class B accuracy but not Class A. As for the mean delivery volume from each pipette, they are 995.7 for pipette A, 986.1 for B and 989.5 for C. Of these, pipette B is just outside Class B accuracy, while both of pipettes A and C are inside the class B tolerance limits.

Thus, whether we classify a result as 'accurate' or 'inaccurate' is determined by the width of pre-established *zones of tolerance*, in just the same way as in target-shooting the diameter of the bull's-eye dictates how many top scores would be secured by a marksman with a particular level of skill and firing from a particular distance.

1.9 Use of y and x

This book departs from the common practice of using x for values of the observational variable and employs y instead. This is done for internal consistency with experiments involving dose–response or analogous relationships (Chapters 7–10). In these more complicated designs, x is the independent and error-free variable, while y is assumed to carry all the random variation. In the pipetting exercise there is no independent variable, therefore no x. But it is still logical to use y as the variable that carries the error.

2
How to Condense
the Bulkiness of Data

I often say that when you can measure what you are speaking about, and express
it in numbers, you know something about it; but when you cannot measure it,
when you cannot express it in numbers, your knowledge is of a meagre and
unsatisfactory kind.

William Thomson [*Lord Kelvin*] *(1824–1907)*

2.1 First, What is Statistics?

2.1.1 Definition of statistics

In the preceding chapter we took a series of 9 readings of the weights of water delivered by 1-cm^3 pipettes and calculated values for the mean ($\bar{y}$) and standard deviation (s). This provided a simple introduction to the science of statistics. What is statistics? It is a branch of applied mathematics which the late Sir Ronald Fisher, one of its great exponents, defined succinctly (1970) as the study of:

(1) *populations*,
(2) *variation*,
(3) *methods for the reduction of data*.

2.1.2 Population and universe

In considering these it must first be pointed out that the word *population* is being used in a special way. It does not necessarily mean a population of *people*. It may be a population of plants, or pipettes or bacteria or mice or atomic particles or, quite generally, *a population or assemblage of observations or measurements*. Allied to *population* is another term, the *universe*, which also has a special usage in statistics different from the normal dictionary meaning of 'All existing things; the whole of Creation (and the Creator)'. To illustrate, we can think of the 9 observations of our pipette experiment as a *random sample* taken from a hypothetical infinitely large collection, or *universe*, of observations which we could *imagine* collecting from pipettes if we repeated the exercise an infinite number of times. However, we could never actually collect the *whole* universe of possible pipette readings because infinity is unattainable. Nevertheless, we can usually obtain sufficiently reliable information (for many practical purposes) about the particular universe in which we happen to be interested simply by accumulating a large enough number of observations. Obviously, we must make sure that the samples are collected in a properly random (or some other kind of unbiased) way so that they are truly representative of the whole.

Returning to *populations*, these can either be of finite size or infinite. For example, a research laboratory may contain a large, but finite, number (population) of 1-cm^3 pipettes—perhaps a few hundred. On the other hand the number of replicate delivery measurements which could be made with any individual pipette is theoretically infinite.

2.1.3 The sample and the population

In statistical thinking and hence in statistical formulae, a sharp conceptual distinction is made between the *sample*, i.e. the actual data collected, and the *population* or *universe*, i.e. the larger (usually hypothetical) aggregate of which the sample is representative. We have used the symbol $\bar{y}$ for the *sample mean*, or *mean-of-the-sample*, and we shall henceforth use the Greek letter μ('mu') when we want to refer to the *population* or *universe mean*. Similarly, we have used s for the standard deviation of the measurements in the *sample* and we shall use σ ('sigma') for the standard deviation of the underlying population or universe. This usage of different symbols is not just an exercise in mathematical hair-splitting, but is a necessary terminology if we wish to understand and describe the relationship between a sample, on the one hand, and the population or universe from which it is drawn, on the other.

In real life the exact values of μ and σ are never, or hardly ever, known with complete exactitude because most of the populations we biologists deal with are infinitely large, and it is impossible to collect an infinite series of measurements. The main use of μ and σ is to provide theoretical reference points when discussing fluctuations in $\bar{y}$ and s in successive samples drawn from the same population. It is rather like an ideal gas in Physics: we need the *concept* of an ideal gas to provide a theoretical basis for the discussion of *real* gases.

2.1.4 Statistical inference

Now we come to the essence of a great deal of statistical work, namely to use observation, or measurements, on *samples* to make inferences about the nature of the underlying *populations* from which the samples were drawn. That is to say, *we observe a sample*, but we *apply the conclusions to a population*. This is an example of the standard scientific procedure of *inductive* reasoning by which we attempt to formulate general conclusions from the observation of specific instances. The opposite procedure of *deductive* reasoning is where we start with a generalization and then predict specific consequences. In science, including statistics, we make use of both.

Thinking again of our pipetting experiment, we should (if we are beginning to see things through the statistician's eyes) regard our found values of $\bar{y} = 20.41$ and $s = 9.7471$ as estimates of the 'underlying' μ and σ values. Just to hammer this message in, let us suppose that you were to repeat the pipetting experiment day after day and to calculate a value of $\bar{y}$ and s for each day. You would find some variation in them from day to day, because of using different pipettes, minor variations in operator skill, changes in barometric pressure (which affects weighing) and all the other little undefinable influences which contribute to experimental variation. So as the days went by you would accumulate a series of slightly different values of $\bar{y}$ and s. In statisticians' language, each day's work would 'provide in $\bar{y}$ and s, an independent estimate of the underlying μ and σ'.

Some people find the above ideas rather disturbing because they conflict with the traditional view of science as a guaranteed method for arriving at the truth. Of course, science *is* concerned with revealing the truths of nature, in so far as it can, but rarely indeed is science able to give *a complete and final* description of anything. Even the full stop at the end of this sentence—the little black dot of ink—is well beyond the capabilities of modern science to describe with complete finality, e.g. in the sense of being able to produce an accurate description of the nature, positions, movements and previous history of all the individual atoms and molecules, electrons, protons, etc. that make up the black dot. So while we can never arrive at the Absolute and Complete Truth of even the simplest thing in nature, we may be able to approximate to it. And some of our approximations may be quite good enough for many purposes. Three final thoughts to consider while we are exploring this philosophical vein.

(1) All processes leading to observations *are* sampling procedures (whether you choose to recognize this or not!). This is one reason why statisticians are fussy about achieving proper randomness in taking a sample—so that the sample may be as representative of the underlying population as it possibly can be.

(2) The numerical values of the *sample mean* and *sample standard deviation* are very unlikely to be exactly the same as the 'true' or population values. Indeed, we could say with little risk of contradiction that they are *never* going to be *exactly* the same, if the population values are defined accurately to an infinite number of decimal places. Therefore in this highly purist sense, the answer you get in a statistical calculation is 'incorrect' or 'wrong' and provides useful ammunition for the snipers who enjoy asserting 'There are three kinds of lies', etc. However it could be argued that on such stringent criteria all other attempts by mankind to arrive at the 'truth' are likely to be even more 'wrong'.

(3) One of the main uses of statistical methods is their power to extract the maximum information from a given set of numerical observations by providing standard methods of summary and judgement. Statistics is therefore nothing more than logic applied to experimental or observational data and formalized by techniques of applied mathematics.

2.1.5 *Parameter and statistic*

There are two other basic terms—parameter and statistic—which should be brought in here. A parameter may be defined as:

(a) a numerical constant which is used to describe a particular population, or

(b) any mathematical function, such as mean and standard deviation, of the measurements from a population.

Mathematically, there is an unlimited number of possible parameters which can be calculated, but some are much more useful than others. We can describe the calculations made so far in the pipette experiment as 'obtaining estimates of the parameters *mean* and *standard deviation* of a universe of weights of pipette deliveries of 1 cm³ of water by a particular student using particular pipettes'.

As for 'statistic' (used in the singular), we can say that the actual numerical value of the arithmetic mean ($\bar{y}$) of our 9 pipette results is *a statistic* which estimates the corresponding parameter value (μ) of the hypothetical universe of pipette deliveries (that exists only as a mathematical idea). Similarly, our calculated value for standard deviation (s) is *a statistic* which estimates the corresponding universe parameter (σ). Thus by applying a mathematical formula to a series of observations in a sample, we 'calculate a statistic in order to estimate the value of a parameter'.

2.2 Constructing a Histogram

2.2.1 Introducing a large mass of data

The above discussion illustrates the importance of the *sample* as a concept in statistics. Now we shall do a simple exercise, still based on the pipetting experiment, in which we shall work with a much larger sample — 135 observations instead of 9. This will illustrate Fisher's third statement: 'statistics is the study of . . . methods for the reduction of data'. As we shall see, such methods, which serve to decrease the bulkiness of a mass of results to a form which the eye and the mind can assimilate, are extremely useful.

For this discussion let us take the results of the pipetting experiment of Chapter 1 as performed by the members of a class of 15 students, each with a

Table 2.1 Results of the pipetting exercise from a class of 15 students. The weights (mg) of water delivered by 1-cm³ pipettes have had the constant 950 subtracted

Student	'Working' weights (mg $-$ 950)								
A	30.2	18.5	39.1	26.7	21.1	15.6	20.8	10.3	6.5
B	57.5	36.8	42.7	31.6	32.0	25.9	47.7	44.7	44.8
C	47.7	50.6	49.7	40.9	49.3	40.1	68.7	45.8	44.8
D	46.3	44.3	50.3	47.2	50.0	82.1	27.1	43.4	52.3
E	82.6	43.6	40.2	38.4	16.7	45.5	62.0	62.7	48.2
F	50.8	49.9	47.6	52.0	44.5	44.7	72.2	70.6	67.2
G	40.3	51.2	24.3	30.7	7.6	0.7	31.3	30.0	30.5
H	87.9	59.8	62.7	85.8	58.0	66.9	72.8	72.9	61.6
I	59.1	63.0	65.7	28.8	32.1	31.3	44.4	49.5	49.6
J	24.1	20.0	19.2	13.3	12.9	22.2	48.7	44.4	53.0
K	40.6	41.3	41.4	45.0	42.0	49.4	41.3	44.2	40.8
L	3.2	23.9	23.9	6.5	34.3	57.8	48.1	48.3	57.1
M	57.2	43.9	61.9	50.3	45.8	25.3	53.5	31.3	42.4
N	51.2	47.6	38.4	54.4	38.0	45.0	76.1	62.7	76.0
O	22.0	41.4	18.0	8.0	8.7	50.9	20.5	8.7	11.0

different set of 3 pipettes. When we put all the results together we have a table with 135 observations (Table 2.1) which is visually indigestible in its raw state. This table shows the 9 results from each of the 15 students. Each value has already been simplified but without loss of accuracy by having a constant of 950 mg subtracted. Take a good look at the table. Does the naked eye allow you to determine what we are going to find (below) when we condense and analyse the data? To anticipate, the statistical treatment will reveal that the 135 measurements are approximately normally distributed, with a mean of 42.18 and a standard deviation of 18.477. Thus the whole mass of 135 measurements can be summarized in a single short sentence. This illustrates how statistical methods can reduce the bulkiness of an assemblage of data and thereby facilitate interpretation.

2.2.2 Assigning data to classes

By casual inspection of Table 2.1 it is virtually impossible to form any impression of what the 135 individual results really show, and obviously some condensation is essential. One simple and useful way of doing this is to construct a histogram, which is known also as a *frequency* diagram or *frequency-distribution* diagram. It shows the relationship between frequency-of-occurrence of particular classes of results and the variable being measured — which here is weight. In this context we are using the word 'class' to mean results which fall within a particular

Table 2.2   Distributing the 135 pipetting results from Table 2.1 into classes of 7-mg width preparatory to plotting a histogram

Upper boundary of class (working mg)	Accumulation of individual measurements from Table 2.1	No. of measurements	% of total
7	////	4	2.963
14	₥ ///	8	5.926
21	₥ ///	8	5.926
28	₥ ₥ /	11	8.148
35	₥ ₥ //	12	8.889
42	₥ ₥ ₥ /	16	11.852
49	₥ ₥ ₥ ₥ ₥ ////	29	21.481
56	₥ ₥ ₥ ////	19	14.074
63	₥ ₥ ////	14	10.370
70	////	4	2.963
77	₥ /	6	4.444
84	//	2	1.481
91	//	2	1.481
		Total 135	99.998[a]

[a]Would be 100.000 but for rounding-off errors in recording the individual percentages to only 3 decimal places.

measurement range such as 0–7 mg on the working scale (equivalent to 950–957 mg on the absolute scale).

In order to plot the histogram, the results first have to be grouped into weight-classes of equal width, e.g. 7 mg, as shown in Table 2.2. All the 135 results in Table 2.1 lie between 0 and 90 mg on the working scale and we can divide them into 13 classes by taking a 7-mg class-width. In fact we can choose any class-width we like (but see Section 2.5.2 below, for the desirable class-width). However to take extremes, a class width of 90 mg would enclose all the results in a single rectangular block (Fig. 2.1a) which would not tell us anything about the shape of the distribution. At the other extreme, a class-width of 2 mg (Fig. 2.1b) tends to give a rather irregular diagram, with discontinuities, which may obscure the general pattern. As a start, one should divide the range of values into 12–15 classes so let us take a class-width of 7 mg and see how it turns out.

In sorting out the raw data into classes, it is convenient to use a table such as Table 2.2 with a wide column in which a 'slash' is made as each result is entered and every fifth slash is made in the reverse direction to form a cluster of 5 for convenience in totalling. Note also that although the class-width is 7 mg, we avoid difficulties in the placement of 'boundary' results such as 21.0 mg by defining the class limit in terms of the upper boundary. Thus the 21.0 mg result is unambiguously placed in the class > 14.00 to 21.0 mg and not in the succeeding class > 21.00 to 28.0 mg. Obviously, each result should be entered only once.

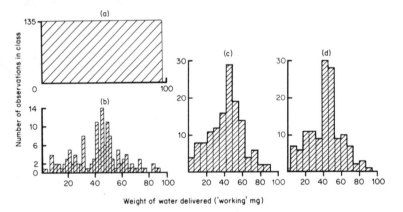

Fig. 2.1 Four histograms of the *same set* of 135 pipetting results to show how the effect of bar width and bar position can drastically alter the appearance of the diagram: a, all results contained in a single bar of width 90 mg — too *little* detail; b, bar width 2 mg — too *much* detail; c, 7-mg-wide bars with upper boundaries at 7, 14, 28, etc. mg; d, as c, except the upper boundaries are at 3, 10, 17, 24, etc. mg

2.2.3 The art-work

Having entered all the results, the number of slashes is counted in column 3 of Table 2.2, while column 4 expresses these numbers in percentages. The numbers in column 3 should be totalled to check that all 135 of the original results have been used (and none used twice!). Inspection of the length of the slash-rows in the completed Table 2.2 makes it clear that we have a definite pattern in the results, with a peak-frequency in the 49 mg class, tailing off on either side to a low value. If we turn the table through 90° we shall, in fact, have the histogram in Fig. 2.1c. As we shall see below, this histogram is only a small part of the use to which Table 2.2 will be put. So Table 2.2 is worth constructing even if we do not bother to draw the histogram.

Note how in Fig. 2.1d the shape of the histogram is quite strongly affected by a 'frame shift' in which the 7-mg-wide class boundaries were chosen at values which are all displaced 3 mg to the right of those used in the adjacent diagram.

2.3 Interpreting the Histogram

A histogram is to be regarded as a useful first summary of a set of observations, provided you have enough of them — say at least 50. When inspecting such a diagram (Fig. 2.1c) a number of points should be considered.

2.3.1 The mode

The highest bar, the 42–49 mg class in Fig. 2.1c, is referred to as the *modal class*, or the class which contains the *mode*. Like the arithmetic mean, the mode belongs to the category of parameters referred to as 'measures of central tendency'. The mode and the mean are usually not numerically identical in a set of experimental measurements although they may be close to each other. They are called 'measures of central tendency' because they refer to that feature often shown by replicate measurements of clustering around a central value. The mode and the mean are only identical if the distribution is fully symmetrical (which this one is not quite). Note that the mode is the highest *point* on the *smooth curve* which can be fitted (Fig. 2.10) to the distribution and is therefore not the tallest *bar* of the histogram. This tallest bar may *contain* the mode but is not itself the mode.

2.3.2 Frequency distribution and the areas within the histogram

The histogram itself is a *frequency diagram* because it shows the frequency-of-occurrence of results within particular classes. For example, the tallest bar, which represents the class from 42.001 to 49.0 mg, contains 29 observations and therefore has a relative frequency of 29/135 or 21.5%. Or, we can express the relative frequency as the decimal 0.215 of 1.000. This leads us into stating what is a most important — indeed, a crucial — idea to grasp.

The area under the *whole* distribution represents 100% of the population of results. In Fig. 2.1c the tallest bar, which contains 21.5% of the population, therefore comprises 21.5% of the area. In Fig. 2.2 the area of the histogram has been divided up and shaded in two ways. In 2.2a the central 7 bars have

24

been shaded so as to divide up the histogram into a central zone which contains a majority of the observations, leaving 'tails' on each side. The central shaded area represents 105 of the 135 observations, which corresponds to 77.8% of the total number. The shaded area therefore makes up 77.8% of the histogram. Outside the shaded area are what are called the lower and upper *tails* of the distribution. These are shaded for prominence in Fig. 2.2b, the lower tail containing 14.8% of the observations and the upper tail 7.4%. This is set out here in some detail to emphasize that the relative area which is vertically under a part, or parts, of the outline of a distribution is a direct measure of the relative frequency-of-occurrence of particular observations. This idea comes up repeatedly throughout this book. All of the *Tests of Significance* (Chapter 3) depend on the idea, so it is of the utmost importance to become familiar with it. Likewise the process of dividing up a frequency-distribution diagram into a broad central zone, with a relatively small tail on each side, is a much-used procedure.

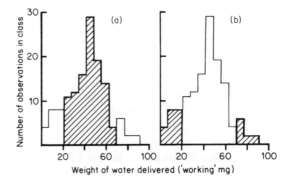

Fig. 2.2 Diagrams to illustrate the important relationship between the area within a histogram and the relative frequencies of particular observations. In (a) the central zone of the histogram, enclosing the 105 pipetting measurements from 21.001 mg to 70 mg is shaded. In (b) attention is focused on the 20 observations between 0 and 21 mg in the lower tail of the distribution, and the 10 observations from 70.001 to 91 mg in the upper tail

2.3.3 Frequency and probability

We saw in the above example that a given class of results, say those lying between 20.001 and 70.0 mg, made up 77.8% of the total. We can say the same thing slightly differently by stating that measurements between 20.001 and 70.0 mg had a relative frequency-of-occurrence (or just 'frequency' for short) of 77.8%. This constitutes a *record of past events* which we can now use for the *prediction of future events*. That is to say, we can use our existing population of 135 observations as a basis for predicting the frequency-of-occurrence of particular individual results if *further* random samples were to be obtained from the *same* population on *another* occasion — we would of course be assuming that the observations were made under the same laboratory conditions as before. We could, for example, predict that if a *single* additional randomly chosen observation were to be added to our existing 135, then there is a probability of 77.8%, or about 3 chances in 4, that it would fall within the range 20.001 to 70.0 mg. Thus, the process of *statistical inference* (with its accompanying assumptions) allows us to equate in a numerical fashion the *frequency of past events* and the *probability of future, or imaginary, events.* This is a subject we return to repeatedly because it is one of the central ideas in statistical methodology.

Philosophically there should be nothing particularly disturbing about these procedures. We do it all the time in everyday life: we base our expectations about future events on our observations of the pattern of similar past events. Statistical inference helps to put a quantitative polish on the framework of these expectations.

2.4 The Normal Distribution

If instead of having only 135 observations as above, we had a very large number of them, say one million, we could imagine plotting a histogram with very narrow bars, but still with large numbers of observations available for inclusion in each bar. If we had an *infinite* number of observations and the bars were made

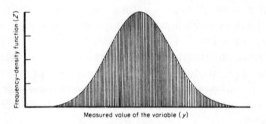
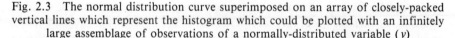

Fig. 2.3   The normal distribution curve superimposed on an array of closely-packed vertical lines which represent the histogram which could be plotted with an infinitely large assemblage of observations of a normally-distributed variable (y)

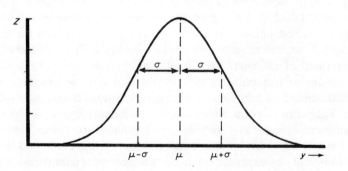

Fig. 2.4   Normal distribution curve: showing μ, the mean of the population of y-values, and σ, the standard deviation, which is the distance from the mean to the points of inflexion on each side of the curve

infinitely narrow, we would finish up with an array of tightly packed vertical lines (Fig. 2.3), whose tops *might* form the outline of a smooth, more or less symmetrical, and more or less bell-shaped curve. *One* such perfectly symmetrical curve, whose shape is illustrated in Fig. 2.3, is known by various names, namely the Normal curve, the Normal Distribution curve and the Gaussian Distribution curve.

The normal distribution curve can be represented algebraically by the formula which may be found in the Addendum to this chapter and which need not divert us here. However we do need to know that μ, the mean, is the value of y that divides the area under the normal curve into 2 symmetrical halves, as shown in Fig. 2.4. The standard deviation, σ, is the horizontal distance on either side of μ which reaches out to the points of inflexion (points where the slope changes over from getting steeper to getting flatter).

2.4.1 Application to experimental measurements

It has been found in many branches of science that the replicate measurements obtained in experiments often follow a normal distribution curve—at least approximately. Only rarely, however, does the chance arise to collect the truly massive number of observations (say, thousands) which would give a histogram with an outline smooth enough to be a really close approximation to the normal curve, as illustrated in Fig. 2.4. Nevertheless, for many purposes, unless there is contrary information, experimental measurements of variables such as weight, length, volume, optical density and temperature can usually be analysed statistically *on the assumption that they represent samples from a normal distribution*, i.e. samples from an underlying normally-distributed universe of measurements.

Theoretically, the normal distribution only applies to what are called *continuous variables*, i.e. measurements like length, weight, etc. which can be made accurate to as many decimal places as you are able. These are distinct from discontinuous variables (i.e. whole numbers), such as the number of dead mice in a cage, or the number of bacterial colonies on a culture plate. You cannot have 4.6 mice killed in a group of 10 injected with a dose of toxin! Despite this theoretical restriction, the normal distribution often *can* be applied to discontinuous, or 'count', data, as we shall see in the chapter on the counting of bacteria and of radioactivity (Chapter 6).

Another theoretical point is that the variable, to be properly normally distributed, should be capable of taking on any value between plus and minus infinity. Again this is not a restriction which should deter us from analysing our experimental results *as if they did fit* a normal distribution. But we should always have in the back of our mind the fact that we *are* making certain assumptions and, as experience with a test system accumulates, we should periodically check the validity of these assumptions. It is quite *unlikely* that our data will be a *perfect fit* to a normal distribution—and it is not necessary that they should! But we should be alert for regular and persistent departures from normality and adjust our statistical procedures if necessary.

One of the prime ways in which statistical analysis can generate false conclusions is when the data are analysed by techniques which are inappropriate to the underlying frequency distribution. This may arise if one makes a wrong assumption about the *shape* of the frequency distribution of the underlying population from which the sample has been taken. It is probably this type of fault, together with biased sampling, which gives statistics the bad reputation which some people enjoy assigning to it.

2.4.2 Areas under the normal curve

Mathematically, it can be shown that the area under the central part of the normal curve enclosed by $\pm \sigma$ on either side of μ is 68.2% of the total area (Fig. 2.5). That is, in a truly normally-distributed population of measurements you would expect to find this percentage of them within these limits. Thus, to anticipate the results of calculations given in Section 2.5: the mean (which we

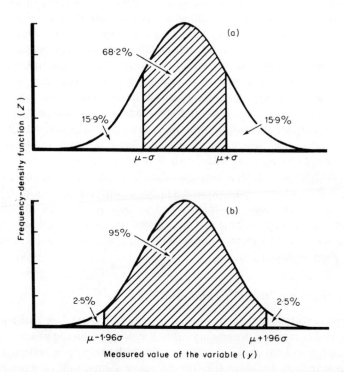

Fig. 2.5 Two common ways of dividing up the area under a normal distribution curve. In (a) the vertical lines are ruled at ± 1 standard deviation (σ) on either side of the mean (μ), while in (b) the rulings are at ± 1.96 standard deviations. In this latter figure, the shaded area makes up 95% of the area under the curve, leaving 5% in the tails, 2.5% in each

can take as μ rather than $\bar{y}$ because our N is so large) of the 135 pipette deliveries in Table 2.1 is 42.1815 mg and the standard deviation (σ) is 18.477 mg. Thus the range of values corresponding to $\mu \pm \sigma$ is from 23.70 to 60.66 mg. If our data are truly normally distributed, then we would expect 68.2% of them, or 92 observations, to fall within these limits. In fact, we find that 91 do, which is very close to the theoretical expectation. Similarly, $\mu \pm 1.96\,\sigma$ is the range 5.97 to 78.40 mg within which we could expect to find 95%, or 128, of our observations. Actually, 129 of them lie within this range, which again is very close to theory. Thus although our histogram in Fig. 2.1c may not look to the naked eye like a normal distribution, we are beginning to accumulate evidence that it may be quite close to normal.

2.4.3 Symmetry of our distribution

Another simple test of normality is to check the *symmetry* of our experimental distribution. A normal distribution is fully symmetrical, and we should therefore expect approximately one-half of our 135 observations (i.e. 67.5 of them) to fall on each side of the mean value 42.18. Actually 76 are above and 59 are below the mean, which suggests that there may be some degree of asymmetry in the distribution. On the other hand, we have to allow for 'random-sampling fluctuations'. That is to say, in a random sample of 135 observations from a genuine normal distribution, we would not necessarily expect to have exactly one-half of them on either side of the mean. It is like tossing a perfect coin 10 times — we do not expect to get exactly 5 heads and 5 tails every time we make a set of 10 tosses. We must therefore apply a test to see if the figures 76 and 59 *depart significantly* from the *expected* value of 67.5. The test in question is the χ^2-test, which is simple to do but is deferred until Chapter 5. Referring ahead to where the calculation is done, (Section 5.7) shows that our figures 76 and 59 do *not* depart *significantly* from the expected value of 67.5. So although there *may* be some skewness in our underlying population of pipette delivery weights, the data are insufficient to establish it. We can therefore proceed on the assumption that any asymmetry in the distribution of our 135 observations about their mean value is slight, and 'within the limits of random-sampling fluctuations'.

2.4.4 Comparing different normal curves

It is important to realize that the normal distribution curve is actually a *family* of bell-shaped curves which can vary (a) in the position on the abscissa of their mean value (μ) and (b) in their degree of 'spreadoutness', as reflected by the size of the standard deviation (σ). Therefore we should cultivate the habit of imagining the normal curve as having the capacity to be shifted horizontally on the abscissa and also to be narrower or more spreadout. Note that we are not here referring to *departures* from normality (see below), but rather to the range of variability that may exist while still firmly within the framework of genuine normal distributions. For example, if we plot the smooth, normal distribution curve with the values of $\mu = 42.1815$ and $\sigma = 18.477$ (the values from the 135 pipetting measurements) we get the curve in Fig. 2.6a. (Note that the abscissa is labelled in actual milligrams.)

Imagine now that our 135 measurements were gathered by a class of students who were capable of much more precise pipetting and who thus achieved a standard deviation one-half of the above, i.e. $\sigma = 9.24$. This would give us the taller and narrower curve shown in Fig. 2.6b — still a genuine normal distribution. Contrariwise, if σ were twice the value of 18.477 we would get the very spreadout distribution in Fig. 2.6c.

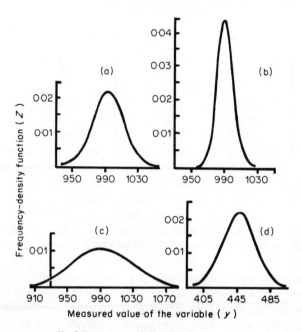

Fig. 2.6   Four curves, all of them normal distribution curves, but varying in their values of μ and σ (see text)

Let us now keep σ constant at 18.477 and vary μ. Suppose we got our class of students to use the same 1-cm³ pipettes but to deliver 0.5 cm³ volumes instead of 1 cm³. This would give us the curve in Fig. 2.6d which has the same degree of spreadoutness as Fig. 2.6a but is shifted to the left.

In conclusion, we must re-emphasize that the different curves shown in Fig. 2.6 are all genuine normal distributions and must not be confused with curves that show *departures* from normality through skewness or kurtosis (see below).

2.4.5   Common departures from normality

Having stated that many types of experimental measurements have been found to be at least approximately normally distributed, while other types are definitely not, perhaps we should deal with the latter before proceeding with further discussion of the normal distribution. A short list of biological examples of *non-normal* distributions would include the following.

(1) The titres of antibody developed in animals given a particular dose of a vaccine or other immunogen (Reid, 1968). The distribution of such titres tends to be markedly asymmetrical, with skewing towards higher-than-average values. Symmetry may however often be achieved by the simple device of plotting the *logarithms of the titres*—in which case the original distribution of the untransformed titres is described as being log-normal (Fig. 2.7). Having made the logarithmic transformation, one then proceeds with the new 'normalized' results as we have done with our pipetting data.

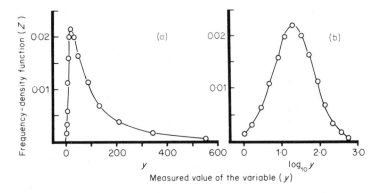

Fig. 2.7   A, a log-normal distribution curve which, by converting the values of y to $\log_{10} y$, gives the symmetrical normal distribution curve in b

If the data *are* log-normally distributed, then the appropriate measure of central tendency is the geometric mean whose formula is given in the Glossary.

(2) Some mammalian physiological data are log-normally distributed, notable examples being insulin levels in human plasma (Welborn *et al.*, 1966) and in mouse serum (Furman *et al.*, 1981), and leucocyte counts in mice that have been injected with endotoxin (Kurokawa *et al.*, 1974) and in humans (Lutz, 1967). On the other hand, blood glucose levels in man and mouse seem to be normally distributed (Welborn *et al.*, 1966; Furman *et al.*, 1981). Epiphytic bacteria on leaf surfaces are log-normally distributed (Hirano *et al.*, 1982).

(3) Again, an animal example, but one where taking logarithms may not help: the survival times of individual animals given a particular dose of a toxin or of an infectious agent. This distribution, especially at the LD_{50} dose, is liable to be highly asymmetrical, with no deaths at all occurring for a time, then a period with many deaths, followed by a long 'tail' which may include some animals that survive indefinitely (or would survive probably to the end of the normal life span, if kept). With such data one can not calculate the mean and standard deviation as we did with the pipette deliveries. Nor can we simply take logarithms, because some of the data may be of the type '> 60 days' and with

no actual value which could be transformed into logarithms. Instead of the arithmetic mean, or the mean logarithm, we would probably take the *median* as the appropriate *measure of central tendency*. The median is defined as the middle value of the variable, when the data are arranged in order of increasing magnitude. This may allow us to utilize information like ' > 60 days' at the upper end of a series.

(5) Bacterial colony counts, virus plaque counts, radioactivity counts, haemocytometer and Coulter counts: these are dealt with in Chapter 6. In fact, provided the counts are large (>30) we can usually treat them as being approximately normally distributed.

(6) Data which consist of whole-number proportions, such as 5/10 for the number of mice killed by a dose of toxin, or 2/100 for the number of contaminated ampoules in a manufacturers' batch of a drug. Again these are dealt with later (Chapter 5) and in certain circumstances can be 'normalized'.

Another type of departure from normality is that known as *kurtosis*. This describes distributions which are symmetrical, but which depart from normality by having either 'too little shoulder' or 'too much shoulder' (see Fig. 2.9b below). Such distributions are respectively more sharply peaked (positive kurtosis) or more extensively flattened (negative kurtosis) than the bell shape of a 'pure' normal distribution. The histograms (Fig. 2.1c and d) of the 135 pipetting results suggest that there is some degree of positive kurtosis in the distribution of these data.

In conclusion, we may note that there is nothing particularly 'normal' about the normal distribution except that it is commonly ('normally') approximated to in many types of experimental measurements. It is another of these statistical terms which has a special meaning different from, although related to, ordinary English usage.

2.5 Estimating the Mean and Standard Deviation of a Large Sample

2.5.1 The tedious way

To calculate the mean and standard deviation of a large sample, such as the 135 pipetting results under discussion, there is a choice of 2 procedures. First, one can simply (but rather laboriously) substitute the data into Eq. 1.3:

$$s = \sqrt{\frac{\Sigma y^2 - (\Sigma y)^2/N}{(N-1)}} \quad \dots\dots\dots\dots\dots\dots\text{Eq. 1.3}$$

as was done with the smaller set of 9 observations in Chapter 1. There is no particular problem about this, except that if the telephone rings when you are half-way through, you may lose your place and make a mistake. It is therefore desirable to do this type of calculation while you are sitting alone on a desert island or using a calculator which gives a print-out of the data entered so that you can check the paper record against the original figures as a safeguard against a simple copying error (which is very likely with as many figures as these). When the measurements of Table 2.1 are substituted into Eq. 1.3 the following results are obtained:

$$\bar{y} = 42.287$$
$$s = 18.497$$

2.5.2 A less laborious method

The above is a very direct procedure for calculating $\bar{y}$ and s, but is rather laborious and can be simplified. An alternative, and much less tedious method now follows. It involves minimal calculation and no entry of very long columns of figures into the calculator. It is especially quick if you have already grouped the results into measurement classes as was done for plotting the histogram, so we shall now capitalize on the effort already made. If you have *not* already done this grouping, then the first step is to prepared Table 2.2. The general name for the procedure is 'calculating the mean and the standard deviation from grouped data'.

Table 2.3   Calculation of mean and standard deviation from grouped data (based on the 135 pipetting results, segregated into 7-mg classes in Table 2.2)

Upper boundary of class (working mg) (1)	Central value of class (Y) (2)	Number of results (f) (3)	Code number for class (U) (4)	$f \cdot U$ (5)	$f \cdot U^2$ (6)
7	3.5	4	−6	−24	144
14	10.5	8	−5	−40	200
21	17.5	8	−4	−32	128
28	24.5	11	−3	−33	99
35	31.5	12	−2	−24	48
42	38.5	16	−1	−16	16
49	45.5	29	0	0	0
56	52.5	19	1	19	19
63	59.5	14	2	28	56
70	66.5	4	3	12	36
77	73.5	6	4	24	96
84	80.5	2	5	10	50
91	87.5	2	6	12	72
Totals		135	0	−64	964

The next step is to prepare a blank version of Table 2.3: for the present, just note the 6 column headings which will be explained shortly. Straightaway we can fill in columns 1 and 3 from the figures in Table 2.2

The second column of Table 2.3 lists the central value (in milligrams) of each class and then we come to column 4 which is the key to understanding the whole table. It involves yet another — but a very simple — transformation of the data. First we pick out the *modal* class, i.e. the group of results containing the most observations. This is the 42–49 mg class which has a centre at 45.5 mg. Now here is the crucial operation: we assign to this class the *code number* zero and then give code numbers 1, 2, 3, etc. stepwise to the classes above and below the modal group, attaching a minus sign to those with smaller milligram values and a plus sign to those with larger milligram values. Our 13 frequency classes are now labelled with code numbers from -6 through zero to $+6$ instead of being labelled with milligrams. We assign the symbols U for the class code number and Y for the central point of the class in milligrams. The group width (7 mg) is given the symbol I.

The relationship between U, Y and I is expressed by:

$$Y = G + IU \quad \dots\dots\dots\dots\dots\dots\dots\dots\dots\dots\dots \text{Eq. 2.1}$$

where G is the milligram value of Y when $U = 0$, i.e. it is the Y-value of the modal group, which here is 45.5 mg.

For column 5 of Table 2.3 we multiply each frequency value in column 3 (f-value) by the corresponding U-value to get $f \cdot U$. We then add up the column of $f \cdot U$ values, taking account of sign. This gives us $\Sigma f \cdot U$ equal to -64. Working in coded units, the arithmetic mean ($\bar{U}$) is given by:

$$\bar{U} = \Sigma f \cdot U / N \dots\dots\dots\dots\dots\dots\dots\dots\dots\dots \text{Eq. 2.2}$$
$$= -64/135$$
$$= -0.47407.$$

To convert this back from the coded scale to 'working milligrams' we use:

$$\bar{Y} = G + I\bar{U},$$

which gives

$$\bar{Y} = 45.5 + 7 \, (-0.47407)$$
$$= 42.1815$$

This is very close (within 0.25%) to the value 42.287 obtained by entering all the 135 results individually into the calculator.

The standard deviation is calculated using the figures in column 6 of Table 2.3, headed $f \cdot U^2$. These values are calculated by squaring each U-value and multiplying by its corresponding f. For example, the first one is $(-6)^2 \times 4 = 144$. The total of this column is $\Sigma f \cdot U^2$ and equals 964.

$\Sigma f \cdot U^2$ is then equivalent to Σy^2 in the standard deviation formula (Eq. 1.3) for ungrouped data. Likewise, $\Sigma f \cdot U$ is equivalent to Σy, and $(\Sigma f \cdot U)^2$ is equivalent to $(\Sigma y)^2$.

In the coded scale, the standard deviation of U, which we can designate s_u, is given by

$$s_u = \sqrt{\frac{\Sigma(f \cdot U^2) - (\Sigma f \cdot U)^2/N}{(N-1)}} \quad \dots\dots\dots\dots\text{Eq. 2.3}$$

Substituting numbers we get

$$s_u = \sqrt{\frac{964 - (-64)^2/135}{134}}$$

$$= \sqrt{\frac{964 - 30.34074}{134}}$$

$$= \sqrt{6.9676}$$

$$= 2.6396$$

To convert the code scale to the milligram scale, multiply by the group width (I). This gives $s = 2.6396 \times 7 = 18.477$ mg which is within 0.11% of the value 18.497 obtained by entering all the individual results into the usual formula for standard deviation. Readers who are following through these calculations on their own calculator may sometimes get answers slightly different, in the last figure, from those given here. This is because small differences can arise between calculations that are done completely on the machine, without writing down intermediate results, and those where the intermediate results are noted down and excessive digits rounded off.

The general rule for using grouped data to estimate mean and standard deviation is that the width of the group should be substantially less than the standard deviation, otherwise a regrouping with a smaller group-width and a repeat calculation should be done. With a 7-mg group-width and a standard deviation of 18.5 we are all right here, whereas a 20-mg group-width would have been much too wide. Ideally, it is recommended that the group-width should not exceed 1/4 of the standard deviation, but only a very large population would justify the extra labour of dividing the data into such a large number of classes.

2.6 Representing the Normal Curve as a Straight Line

To represent the bell-shaped normal distribution curve as a straight line may seem a rather strange thing to want to do. But as we shall see, it is a worthwhile procedure for helping to demonstrate the normality or otherwise of a set of data such as our 135 pipetting results. The underlying reason is that the human eye finds it easier to detect the departure of experimental points from a theoretical straight line than to assess departures of points from a curve.

Therefore if we transform the bell-shaped normal distribution curve into a straight line by a simple mathematical device, we can then plot our experimental results on top and see how closely they fit. If the data *are* normally distributed, or approximately so, they will be randomly scattered closely around the theoretical straight line given by a perfect normal distribution. The procedure is straightforward and makes use of the effort already expended in grouping the 135 pipetting results into 7-mg wide classes. It also uses the values of mean ($\bar{y}$ or μ) = 42.1815) and standard deviation (s or σ) = 18.477 calculated in the previous section.

2.6.1 Probit transformation

The device for changing the bell-shaped normal curve into a straight line is known as the probit transformation and is done by means of tables (Appendix A.6). It is not necessary to understand probits in order to use them, any more than one needs to know how to calculate π from first principles to be able to calculate the area of a circle as πr^2. However, it *is* useful to be aware that probits exist on a scale that goes from about 2 to about 8 and have units equal to the standard deviation of the system under investigation. Also, one should know that the number 5 is the middle of the probit scale and corresponds to the arithmetic mean of the population being studied. Thus, a probit value of 4.0 corresponds to a distance of 1.0 standard deviation below the mean, and a probit of 3.0 corresponds to 2.0 standard deviations below the mean. The reason the probit scale does not go much below 2 is that this value corresponds to a distance of 3 standard deviations below the mean, which is seldom reached by experimental data.

Because the probit scale is symmetrical, the probit values of 6.0, 7.0 and 8.0 likewise correspond to distances of 1.0, 2.0 and 3.0 standard deviations *above* the mean.

Let us now apply the probit technique to our 135 pipetting results. We have already calculated the overall mean $\bar{y} = 42.1815$ and $s = 18.477$. We now calculate:

$$\bar{y} - 2s = 42.1815 - 2 \times 18.477$$
$$= 5.23 \text{ mg}$$

and

$$\bar{y} + 2s = 42.1815 + 2 \times 18.477$$
$$= 79.14 \text{ mg}$$

The first of these corresponds to a probit value of 3.0 and the second to a probit value of 7.0. A graph is then plotted of probit, on the ordinate, against weight (mg) on the abscissa. The position of the straight line is defined by the points (5.23 mg, probit = 3.0) and (79.14 mg, probit = 7.0), as shown in Fig. 2.8a. This graph is thus a straight-line representation of the bell-shaped normal distribution curve with $\mu = 42.1815$ and $\sigma = 18.477$.

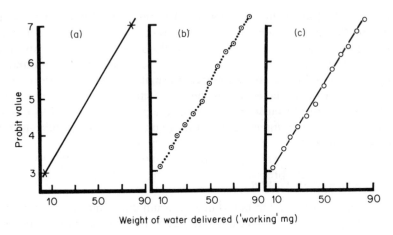

Fig. 2.8 Probit plots from the pipetting experiment: a, the straight line probit plot that represents the normal distribution curve with $\mu = 42.1815$ and $\sigma = 18.477$; b, the experimental probit plot of the 135 pipetting measurements as in Table 2.4; c, result of superimposing a and b

Table 2.4 Probits corresponding to the accumulated percentages in the 7-mg groups of the 135 pipetting results

Upper boundary of the 7-mg class (mg) (1)	Number of results in class (2)	Accumulated number of results at upper boundary of class (3)	Accumulated percentage at upper boundary of class (4)	Probit of accumulated percentage (5)
7	4	4	3.0	3.12
14	8	12	8.9	3.65
21	8	20	14.8	3.95
28	11	31	23.0	4.26
35	12	43	31.9	4.53
42	16	59	43.7	4.84
49	29	88	65.2	5.39
56	19	107	79.3	5.82
63	14	121	89.6	6.26
70	4	125	92.6	6.45
77	6	131	97.0	6.88
84	2	133	98.5	7.17
91	2	135	100.0	—

2.6.2 Probits of the experimental results

The probit values for the experimental data are determined from the tables in which the 135 results were earlier grouped into 7-mg-wide classes for plotting the histograms (Table 2.2) and for calculating the mean and standard deviation by the grouping procedure (Table 2.3). For convenience Table 2.4 presents in its first two columns the upper boundaries of the 7-mg-wide grouping classes and the number of results in each class. Column 3 is derived from column 2 by successive addition of the number of results in each class, so as to get the *accumulated* number of results, from the bottom of the distribution up to the upper boundary of each class. Note that at the base of this column the entry is 135, indicative of all the results having been entered.

Column 4 expresses each column 3 value as a percentage of 135 to give the percentage of the distribution which has been accumulated at the upper boundary of each grouping class. The probits in the final column of the table are obtained by entering each percentage figure in column 4 into the probit table (Appendix A6) and reading out the probit value. There is no probit value for 100% which is why the last entry in column 5 is left blank.

The final step is to plot a graph of probit (ordinate) against upper boundary of each class in milligrams, on the abscissa, as shown in Fig. 2.8b. The successive points are then joined up and yield a slightly zig-zag line. The question of how closely this experimental line approximates to the theoretical straight line of a perfect normal distribution is answered in Fig. 2.8c, where the two plots are superimposed. It is clear that the fit is good and that the experimental line wanders around the theoretical line in a seemingly random fashion and with no trend for systematic departures of the type seen below when we come to deal with definitely non-normal distributions.

Further analysis of the goodness-of-fit of the experimental points around the straight line is given in Chapter 5, Section 5.7.

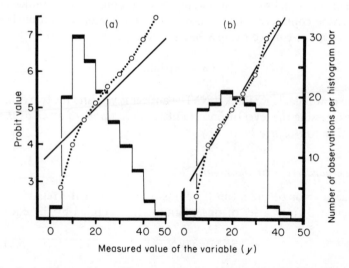

Fig. 2.9 Composite probit and frequency diagrams for two non-normal distributions. Superimposed on each histogram is the theoretical straight line of probit versus y-value calculated as described in Section 2.6.1, together with the experimental probit plot (dotted line) calculated by the method in Table 2.4. (a) is a skewed distribution of $N = 139$ observations with $\mu = 20.09$ and $\sigma = 15.39$; (b) is a distribution exhibiting negative kurtosis, i.e. flat-topped, with $N = 119$, $\mu = 20.31$ and $\sigma = 9.06$

2.6.3   Effect of skewness and kurtosis

An example of data that follow a skewed distribution is given in Fig. 2.9a, where on the same diagram we have a histogram, a probit plot and the independently-positioned straight line, as done in Fig. 2.8. It is clear that the observations summarized here are *not* a good approximation to a normal distribution.

Another type of departure from normality is illustrated in Fig. 2.9b, where the histogram exhibits *negative kurtosis*. Here the top of the distribution is flattened more than it should be for true normality. The probit method shows this departure from normality by systematic (rather than random) 'wandering' of the experimental probits away from the theoretical normal line.

Both skewness and kurtosis can be investigated quantitatively and the degree of each expressed numerically as coefficients of skewness or kurtosis. The calculations are not given here but may be found in Snedecor and Cochran (1967, p. 86). These authors also provide methods whereby a given distribution may be analysed to see if the degrees of skewness or kurtosis are statistically significant, or within the limits of sampling fluctuations.

2.7   Uncertainties in Estimating a Mean

It was pointed out earlier in this chapter (Section 2.1.3) that whenever we collect a set of replicate measurements, like our 135 pipetting results, the value we get for $\bar{y}$, the mean, is an *estimate* of an 'underlying' population parameter μ. The question to be considered now is how much error or uncertainty should be attached to $\bar{y}$, i.e. can we express our result as:

$$\bar{y} \pm \text{something}$$

to indicate how reliable we think it is? The answer is yes, and the two commonest ways of expressing the level of uncertainty are *the standard error of the mean* and the *95% confidence limits*.

2.7.1   Standard error of the mean

The standard error of the mean (SEM) is a measure of the reliability of the mean calculated from a set of observations. It is defined by the equation:

$$\text{SEM} = s/\sqrt{N}. \dots\dots\dots\dots\dots\dots\dots\text{Eq. 2.4}$$

where s is the standard deviation and N is the number of observations. With our 135 pipetting observations:

$$\text{SEM} = 18.477 \div \sqrt{135}$$
$$= 1.5902$$

We can therefore express our mean $\pm$ SEM (and with the number of observations) as:

$$42.1815 \pm 1.5902 \ (135)$$

which is a useful way to summarize a set of measurements.

One shortcoming of the SEM is that it does not actually tell you how close the value of $\bar{y}$ is to μ. Nor do the plus and minus values represent extreme limits within which $\bar{y}$ is bound to lie. So what does the SEM represent? Probably the best way to regard it is *an estimate of the standard deviation of $\bar{y}$*. In other words, if you were to repeat, day after day and under the same conditions, the experiment which yielded the 135 observations and calculate a value of $\bar{y}$ for each day, then on 68.2% of a long series of days, the value of $\bar{y}$ could be expected to lie within the limits of 42.1815 ± 1.5902. The virtue of the SEM is that it allows you to make this prediction of variability from the results of a *single set of observations*.

It is important to keep clear the difference between s, the standard deviation, and $s/\sqrt{N}$, which is the SEM. Thus, s is a measure of the intrinsic variability, or degree of scatter, of *individual observations* in the sample. As we take more and more observations (i.e. N increases), s becomes a closer and closer approximation to σ, the standard deviation of the hypothetical underlying population. In the limit, when $N = \infty$, $s = \sigma$. With SEM, on the other hand, as N approaches ∞, SEM *approaches zero*, because of the position of N in the divisor of the SEM formula. Thus, as N *reaches* ∞, we can express our mean and its degree of uncertainty (SEM) as $\bar{y} \pm 0$. But during this process s is *not* approaching zero, but simply becoming indistinguishable from σ.

2.7.2 95% Confidence limits

When we calculate the $\bar{y}$-value of a set of observations, we are making what is called a *point estimate* of the underlying population value μ. Alternatively, as in what follows, we could make an *interval estimate* of μ by calculating lower and upper boundary values within which μ has a high probability of being bracketed or enclosed. The 95% confidence limits (95% CL) are bracket values of this kind and are calculated by the formula:

$$95\% \ CL = \bar{y} \pm t(SEM) \dots\dots\dots\dots\dots\dots Eq. \ 2.5$$

which is exactly the same as Eq. 2.4 used in the previous section, except for the symbol t. This is the 'Student t-statistic' and is found by looking up the t-table in the Appendix (Appendix A2). (The t-test is discussed extensively in the next chapter and therefore gets only brief mention here.)

Referring to the t-table, we focus attention on the column with 5 at the top and note that the values in the column—these are values of t—decrease steadily as we go down. Next we need to know that 'degrees of freedom' in the present context are simply the *number of observations minus 1*. So with $N = 135$ observations, we have 134 degrees of freedom. The table does not have an entry for this actual number in its left-hand column but this does not matter greatly, as t only changes from 1.98 to 1.96 when one goes from 120 to ∞ degrees of freedom. So we can interpolate 134 degrees of freedom as having a t-value of about 1.97.

Substituting into Eq. 2.5 gives us;

$$95\% \ CL = 42.1815 \pm 1.97 \ (1.5902)$$
$$= 39.04 \text{ and } 45.31$$

These confidence limits allow us to state that 'μ (the underlying population parameter of which $\bar{y}$ is an estimate) has a 95% chance of lying within the limits of 39.04 and 45.31 mg'. We cannot say *where* μ lies within these limits. Nor can we say that μ *definitely* lies within these limits. However, we *can* make the *probability statement* that 'μ has a 95% chance of being *somewhere* within the 95% CL'. Thus, to recapitulate: our value $\bar{y}$ is a *point* estimate of μ, while the 95% CL provide an *interval* estimate. If we do larger and larger experiments, so that N approach ∞, the 95% CL get progressively narrower until when $N = \infty$ they disappear altogether and $\bar{y}$ becomes identical with μ.

2.8 Addendum: Plotting a Normal Distribution Curve

For the purposes of this book it is not actually necessary either to know the algebraic formula for the normal distribution or to be able to plot the curve by substituting particular values of μ and σ. However, some readers may like to practice their skills with the pocket calculator or microcomputer and this addendum is therefore provided to round off the chapter.

The equation for the normal distribution is:

$$Z = \frac{1}{\sigma\sqrt{2\pi}} \ e^{-\frac{1}{2}[(y-\mu)/\sigma]^2} \quad \ldots\ldots\ldots\ldots\text{Eq. 2.6}$$

The equation has the two variables, y and Z, and four constants, π, e, σ and μ. The first two constants are the absolute mathematical constants, $\pi = 3.14159$ and $e = 2.71828$. The other two constants, σ and μ, are the *parameters* which are constant for a particular population of measurements, but which vary from one population to another. As already defined, μ, is the parameter which we estimate when we calculate the $\bar{y}$ of a set of measurements, and σ is the parameter estimated by s, the standard deviation. Pictorially, μ and σ can be inserted on the plot of the normal curve, as in Fig. 2.4.

To plot the normal curve we make use of Eq. 2.6, inserting into it our value $\bar{y} = 42.1815$ as an estimate of μ, and $s = 18.477$ as our estimate of σ, both from the grouped-data procedure. Although perhaps slightly intimidating to the

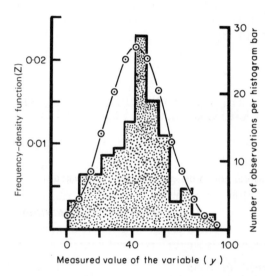

Fig. 2.10 Superimposed plot of the smooth normal distribution curve (calculated in Table 2.5) with parameters $\mu = 42.1815$ and $\sigma = 18.477$ on to the histogram from Fig. 2.1c. Note that the two ordinate scales are different

Table 2.5 Table of points for plotting the smooth curve corresponding to our large sample of 135 pipetting results. It is done by taking a series of suitably spaced values of y (based on the histogram 7-mg boundaries) and calculating the value of the ordinate Z, from Eq. 2.6, substituting $\mu = \bar{y} = 42.1815$ and $\sigma = s = 18.477$

y	Z
0	0.0016
7	0.0035
14	0.0067
21	0.0112
28	0.0161
35	0.0200
42	0.0216
49	0.0202
56	0.0163
63	0.0114
70	0.0069
77	0.0037
84	0.0017
91	0.0007

non-mathematician, Eq. 2.6 is easy to work with on a calculator with buttons for e^x *and* π and with 3 memories. With these features you first calculate $1/\sigma\sqrt{2\pi}$ and store it in memory 1, while the values of μ and σ are stored in memories 2 and 3. Then insert a series of suitably spaced y-values, such as those in Table 2.5, and calculate the corresponding value of Z. The values of y chosen for insertion were those at the 7 mg boundaries as used for grouping. The whole job should take only about 20 minutes. Fig. 2.10 shows the end-result, with the previously obtained histogram (Fig. 2.1c) superimposed. In order to get good superimposition, the two ordinate scales were adjusted (note that they are different) so as to give diagrams of similar size. Otherwise we would have got a small histogram over-shadowed by a large normal curve, or vice versa.

3
Is that Difference Significant?

Read not to contradict and confute, nor to believe and take for granted, nor to find talk and discourse, but to weigh and consider.

Francis Bacon (1561–1626).

3.1 Various Kinds of 'Difference' Problem

Working as an experimental biologist, you will inevitably collect groups of observations which have to be compared with each other to decide whether the differences between them are *significant*. Naturally, you do not need statistics to tell you that a set of culture plates with confluent growth has significantly more microbes than a set that is sterile. But often you may have to make decisions where the differences are much less, and where it is therefore desirable to have some standard procedures for making objective assessments. Incidentally, the word *significant* which we shall see frequently in this chapter has a special meaning in statistics which is not the same as in everyday English.

*Dash it all! 'Good' and 'Evil' MUST be
significantly different!!*
(Courtesy of *Trends in Biochemical Sciences* and
Elsevier Science Publishers)

You will hear some people say that the only kind of experiments they do (and by implication the only kind worth doing!) are those that give such unambiguous data as to render statistical analysis unnecessary. Unfortunately, however, not all of us work with systems that give such clear-cut results all the time, with the consequence that some kind of statistical treatment is often unavoidable. Also, with statistical analysis so commonplace, it is becoming increasingly quaint merely to assess results 'by eye'.

There is a considerable range of 'difference' problems which the biologist may encounter:

(1) Comparing the means and standard deviations of several groups of replicate measurements to see if the differences between the groups are significant.

44

(2) Comparing the survival ratios of groups of animals or humans exposed to various toxic or pathogenic agents to see if the differences are significant, e.g. is 9/20 significantly different from 13/20? (Answer: No it isn't.)

(3) Assessing whether sets of graph points differ significantly from a superimposed straight line.

These are just a few of the 'difference' problems which will be considered in this book. The question now is where to start. There is no single statistical *Test of Significance* that has universal applicability. Instead, there is a whole array of tests, each designed for a particular set of circumstances.

However, despite the apparent diversity of the various Tests of Significance (*t*-test, *F*-test, χ^2-test, *U*-test, etc.), they all work on the same basic set of ideas. So if you understand one test, you will be able to do the others by simple extension. Thus, with one test properly grasped, it is then a matter of knowing the circumstances in which you should or should not use it, and which other tests might be more appropriate. With this approach, our procedure will therefore be:

(1) To indicate which statistical Tests of Significance should be used with which kinds of data, and with which questions being asked about the data, but without explaining the tests in any way, i.e. just giving their names and then signposting you to the sections where they are described (Fig. 3.1).

(2) To describe in detail one of the commonly used Tests of Significance — the *t*-test — and to use it as a base to explain the general philosophy behind Tests

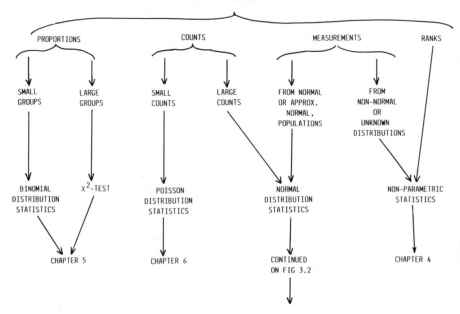

Fig. 3.1 The 4 basic categories to which data belong and signposts to the various tests that are used for their analysis

of Significance in general, as well as specifically how to use the t-test, with examples, precautions, restrictions, extensions, etc.

(3) To deal with the other Tests of Significance systematically, but more in a 'cookery-book' way, and without restating the general philosophy.

3.2 Choosing the Right Statistical Test

The first step is to determine to which of 4 basic categories your data belong. These 4 categories are: Proportions, Counts, Measurements, and Scores or Ranks. Let us take some examples.

3.2.1 Proportions

Here we are thinking of results such as 3/8 for the proportion of broth tubes showing turbidity out of 8 inoculated and incubated; 7/12 for the proportion of mice dead after injection of a toxic or pathogenic agent, or for the proportion of white and brown mice in a litter; 351/14,276 for the proportion of people in the field trial of a vaccine who contract the disease. The essential feature of these examples is that the raw data consist of whole-number proportions which are *discontinuous*, e.g. you cannot have 7.3 mice dead out of 12 inoculated. So although you can express the 7/12 as a decimal 0.5833 (or 58.33%), the raw data themselves consist intrinsically of whole-number proportions. To analyse such data for significance of differences, you should apply tests based on the Binomial Distribution, or use the χ^2-test (Chapter 5).

3.2.2 Counts

These again will be whole numbers (without a decimal) and will emerge from such operations as bacterial colony counts, cell counts in a haemocytometer or Coulter Counter, virus plaque counts and radioactivity counts. Such data resemble the *proportions* in consisting intrinsically of whole numbers without a decimal, but differ in not having a divisor. To deal with these, see Chapter 6.

3.2.3 Measurements

Here the experimental variable is *continuous*, in the sense of not being restricted to whole number values, and the results may be gathered with as many decimal places as the measuring instrument allows. Such data will typically be expressed in units of length, height, area, volume, weight, concentration, temperature or optical density. Measurement data are conceptually distinct from *proportions* or *counts*, although one should not be too rigid in placing boundaries, because counts and measurements can merge into each other. Thus, in weighing, for example, deliveries of 1 cm^3 of water from a 1-cm^3 pipette, the weight can be expressed to as many decimal places as the balance allows, i.e. you are not restricted to whole-number values of weight-units until you get down to the molecular level and express the data as so many indivisible molecules of H_2O (which is obviously impossible in practice). But at a more realistic level, you may be using a balance which reads only to the nearest 0.1 mg. It then becomes an exercise in semantics whether you treat 992.5 mg as a 'measurement of weight' or a 'count of tenths of a milligram'.

3.2.4 Scores, ranks and qualitative assessments

These may not come up very often in biological laboratory work, but you could find yourself dealing with them in serum agglutination tests and pathological reactions of tissues. An example of the former is where titres are scored as $+ + + +, + + +, + +, \pm$, etc. Or the degree of pathology in animals or plants exposed to a noxious agent may be scored as 4, 3, 2, 1, 0 as judged by naked-eye inspection. See Chapter 4 for further discussion.

3.3   Taking the Road to the *t*-Test

The *t*-test is quite specialized in its application and the following criteria have to be fulfilled:

(1) The underlying distribution of the data should be normal, or approximately normal, or 'normalizable' by a transformation.

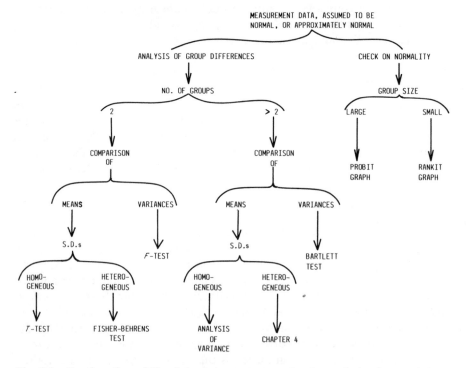

Fig. 3.2   Continuation of Fig. 3.1: signposts to tests for the analysis of normal data

(2) The number of groups to be compared = 2 (this can be extended to a larger number of groups, with modification).

(3) The standard deviations of the 2 groups should not be significantly different.

(4) The purpose of the t-test is to assess the significance of the difference in the means of the 2 groups.

Although in Figs. 3.1 and 3.2 the t-test is only one of what may seem to be a bewildering variety of possibilities, this does not indicate its true importance. The experimental biologist who generates measurement data will use the t-test frequently. So let us take a specific example.

3.3.1 A second pipetting experiment

Although we have not yet exhausted all the possibilities for analysis of the first pipetting experiment of Chapters 1 and 2 (more follow in Chapters 4 and 7) let us now embark on a second pipetting experiment to generate some data on which we can do a t-test. This time let us take a single 1-cm^3 pipette and make 8 deliveries of 1 cm^3 water into a tared weighing vessel. Four of the deliveries are to be made by simply allowing the pipette to drain by touching its tip against the vessel walls, while the other 4 are to be made similarly but with the additional blowing out of the last drop. We shall refer to the two methods as 'drainage' and 'blow-out', respectively, and shall collect them by a random scheme as

Table 3.1 Weights of water obtained from a 1-cm^3 pipette by simple drainage (D), and by drainage plus blowing out (B) the last drop. The measurements were collected in the random sequence DDBDBDBB from the bottom of the table upwards

Method of delivery[a]	Balance reading (g)	Weight of water delivered		
		in grams	in milligrams	'working' number of mg[b]
B	24.1070	1.0148	1014.8	25.8
B	23.0922	1.0109	1010.9	21.9
D	22.0813	0.9971	997.1	8.1
B	21.0842	1.0232	1023.2	34.2
D	20.0610	0.9961	996.1	7.1
B	19.0649	1.0135	1013.5	24.5
D	18.0514	0.9926	992.6	3.6
D	17.0588	0.9896	989.6	0.6
Tare	16.0692			

[a]To randomize the sequence of collecting the 'drainage' and 'blow-out' readings, the former were assigned random digits 1, 2, 3 and 4, and the latter 5, 6, 7 and 8. Reading along the first row of the table of random digits (Table A1) then gave the following assignments:

```
2   0   1   7   4   2   2   8   2   3   1   7   5   9   6
D   —   D   B   D   —   —   B   —   D   —   —   B   —   B
```

where blanks indicate that the number is either not wanted or has already been allocated.
[b]Obtained by subtracting 989.0.

shown in Table 3.1. The layout of this table is very similar to that used in the first pipetting experiment (Chapter 1) and we make the same computational simplifications so as to finish in the right-hand column with a set of data expressed in 'working milligrams'. It is evident to the naked eye, as you would expect, that more water is delivered by 'blow-out' than by 'drainage' and it will therefore be interesting to see how firmly statistical analysis confirms this. So we shall put the question: 'Is there a significant difference in the average amounts of water delivered by "drainage" and by "blow-out"?'

3.3.2 Preliminary processing of the data

Before calculating t, we need the mean and standard deviation of each of the 2 groups of measurements. Table 3.2 unscrambles the randomization and sets out the data obtained by 'drainage' and 'blow-out'. It then provides the values of the means ($\bar{y}_1$ and $\bar{y}_2$) and the standard deviations (s_1 and s_2) but without giving the intermediate steps in the calculations.

Table 3.2   Systematic rearrangement of the data from Table 3.1 and calculation of means and standard deviations

1.   *Tabulation of the 'working' number of milligrams*

Drainage (y_1)	Blow-out (y_2)
0.6	24.5
3.6	34.2
7.1	21.9
8.1	25.8

2.   *Calculation of means*
from $\bar{y} = \Sigma y / N$ . . . . Eq. 1.1
mean of the drainage values ($\bar{y}_1$) = 4.85
mean of the blow-out values ($\bar{y}_2$) = 26.60

3.   *Calculation of the standard deviations*

from $s = \sqrt{\dfrac{\Sigma y^2 - (\Sigma y)^2 / N}{(N-1)}}$ . . . . Eq. 1.3

standard deviation of the drainage values (s_1) = 3.428
standard deviation of the blow-out values (s_2) = 5.320

4.   Note that if the 2 groups have different N-values, n_1 and n_2, and Eq. 3.2 is being used to calculate t, the required value of s for that equation if given by:

$$s = \sqrt{\frac{\Sigma y_1^2 - (\Sigma y_1)^2/n_1 + \Sigma y_2^2 - (\Sigma y_2)^2/n_2}{n_1 + n_2 - 2}}$$

where Σy_1^2 is the sum of squares of the individual measurements in Group 1 and Σy_2^2 the corresponding sum for Group 2

3.3.3 The t-test formulae

Modesty is not one of the prevailing idioms among scientists, and few people would publish a classic paper under a pseudonym. However, in 1908, W. S. Gosset who was a chemist working in the Guinness Brewery in Dublin, published under the authorship of 'Student' an important paper entitled 'The Probable Error of a Mean'. This paper is the basis of what is now known as 'Student's t-test' which is one of the most widely used of the Tests of Significance. The statistic t can be calculated in various ways, one of which is by the following equation:

$$t = \frac{(\bar{y}_2 - \bar{y}_1)\sqrt{N}}{\sqrt{(s_1^2 + s_2^2)}} \dots\dots\dots\dots\dots\dots\dots\dots \text{Eq. 3.1}$$

where $N=$ the number of observations in each group.

Substituting our already available values for $\bar{y}$ and s from Table 3.2 gives

$$t = \frac{(26.60 - 4.85)\sqrt{4}}{\sqrt{(3.428^2 + 5.320^2)}}$$

$$t = 6.873$$

Before we consider how this result is to be interpreted, it would probably be logical to set down the somewhat more complex version of Eq. 3.1 which has to be used if the numbers of observations in the 2 groups are different, e.g. if instead of having N in each group we had n_1 and n_2 in 'delivery' and 'blow-out', respectively, then the t-test equation would be:

$$t = \frac{(\bar{y}_2 - \bar{y}_1)}{s\sqrt{\left(\dfrac{1}{n_1} + \dfrac{1}{n_2}\right)}} \dots\dots\dots\dots\dots\dots \text{Eq. 3.2}$$

which simplifies to Eq. 3.1 if $n_1 = n_2$ and s is calculated as in Table 3.2. One might make the general comment here that statistical formulae tend to be more complicated if they have to cope with asymmetrically designed experiments. Wherever possible, therefore, you should try to have the same number of observations in each group.

There is yet another version of the t-test formula that is used, this time with *paired* data. For example, in the present pipetting experiment we might, instead of having collected the drainage and blow-out deliveries in a purely random sequence, have collected them in 4 pairs, each made up of 1 drainage delivery and 1 blow-out. Let us suppose for argument that we got exactly the same numerical values as those now under consideration and that the data were as in Table 3.3 which shows one of the possible arrangements of paired relationships of the different readings. Note that the pairing as recorded in this table is only

Table 3.3 Example of a *t*-test on paired data (based on Table 3.2)

Pair no.	1	2	3	4
Drainage (y_1)	0.6	3.6	7.1	8.1
Blow-out (y_2)	24.5	34.2	21.9	25.8
Difference ($y_1 - y_2$) (given the new symbol, h)	− 23.9	− 30.6	− 14.8	− 17.7

one of several possible patterns, e.g. instead of having 0.6 paired with 24.5, we could have chosen 8.1 to be paired with 24.5 (or had we been doing the experiment in reality, it might have worked out that way).

However, granted that we are thinking of the data as having been collected in pairs, we proceed to calculate the *difference* between each pair. This difference is then treated as a new variable which, for convenience, we can call h. The values of h are given in the bottom line of Table 3.3

Next is calculated the mean of h, $\bar{h} = -21.75$, and the standard deviation of h, s_h, using Eq. 1.3. This gives $s_h = 7.0155$.

The appropriate *t*-test equation is now:

$$ t = \frac{\bar{h}\sqrt{N}}{s_h} \qquad\dots\dots\dots\dots\dots\dots\dots\dots\dots\text{Eq. 3.3} $$

where N = the number of pairs = 4. This gives:

$$ t = \frac{-21.75\sqrt{4}}{7.0155} = -6.201 $$

which you will note (ignore the minus sign meanwhile) is not numerically the same as the result we got by substituting the same numbers (but without the assumption of pairing) into the first *t*-equation. This re-emphasizes one of the central themes of this book: that the statistical approach has to be built into the experiment before the bench work is done; and the appropriate method for analysing the data will depend on exactly how the data were gathered. And gathering data in pairs is not the same as gathering the same data in a random sequence—even if the actual numerical values are identical.

However, the main purpose of this section was to introduce you to the three most commonly-used equations for calculating the *t*-statistic from grouped or paired data. Let us now interpret the values of t so obtained.

3.3.4 Interpreting a 'found value' of t

The values of t we got by substituting our experimental data into the formulae are known as 'found values' of t. In order to interpret them we have to compare them with the 'tabulated values' of t from the t-table in Appendix A2.

But in order to extract the appropriate values from the t-table for comparison, first we have to determine the number of *degrees of freedom* (d.f.) associated with our data. The rule for this is that if you are comparing two groups with n_1 and n_2 observations, then the number of d.f. is $(n_1 + n_2 - 2)$. So in our case with the randomized data it is $(4 + 4 - 2) = 6$. With paired data the number of d.f. is $(N - 1)$, where N is the number of pairs. Therefore in the calculation where we assumed that the observations had been collected as 4 pairs, the d.f. = 3. Another common abbreviation for degrees of freedom is ϕ. For further discussion of degrees of freedom see the Glossary.

Knowing the d.f., we consult the t-table (Appendix A2), where we find the following entries for 3 and 6 d.f.:

Degrees of freedom	Value of t corresponding to P-value (%)				
	10	5	2	1	0.1
3	2.353	3.182	4.541	5.841	12.924
6	1.943	2.447	3.143	3.707	5.959

P stands for probability, expressed as %. It might equally well have been expressed as a decimal of 1.0, so that 0.1 would correspond to 10% probability, 0.01 to 1% and 0.001 to 0.1%. You must get accustomed to thinking of these probabilities either as percentages or as decimal fractions of 1.0, because both are widely used. Note that P gets smaller as we move across the table from left to right, while the corresponding t-values become progressively larger. Thus, a large t is associated with a small P, and vice versa. The question of what the probability refers to is temporarily shelved until the next section.

Table 3.4   Division of the t-table (shown in abbreviated form) into significance zones

Degrees of freedom	Value of t corresponding to P-value (%)				
	10	5	2	1	0.1
				found value 6.201 ↓	
3	2.353	3.182	4.541	5.841	12.924
6	1.943	2.447	3.143	3.707	5.959
					↑ found value 6.873

Difference in means	Not significant	Significant	Highly significant

Meanwhile for purposes of interpreting 'found' values of t, you need to know that the t-table is divided into 3 zones, with boundaries extending vertically from $P = 5\%$ and 1%. From left to right these vertical zones are labelled *not significant*, *significant* and *highly significant* as shown in Table 3.4. This table also shows our found values, from the examples being worked, interpolated into the 3 d.f. and 6 d.f. lines, as appropriate. Both of our found values of t fall into the 'highly significant' zone, that from the paired data being between $P = 1\%$ and 0.1% and that from the random data being so high as not to lie within the scope of the table which only tabulates down to $P = 0.1\%$. We refer to this latter result as 'very highly significant'.

The terms *not significant*, *significant* and *highly significant* relate to the differences in average amounts of water delivered by 'drainage' and by 'blow-out'. We can therefore summarize the results of our statistical analysis so far by stating: 'Application of the t-test to the randomly-collected blow-out and drainage deliveries shows a very highly significant difference ($P < 0.1\%$) in the mean volumes of water delivered. In the version of the experiment where we imagined the results to have been collected pairwise, the difference is still highly significant ($1\% > P > 0.1\%$) but less so than in the random experiment.'

Referring back to the first paragraph of this chapter, you will appreciate why I did not attempt to define the word 'significant', in its statistical usage, at that early stage in the discussion. Even now, we are not able to give a very neat definition, except something along the lines of: 'The difference in the means of two groups of measurements is *statistically significant* if, after substituting the measurements into the t-equation, you get a *found value* of t which, when interpolated into the t-table with the appropriate number of degrees of freedom, corresponds to a P-value that is less than or equal to 5% but greater than 1%.' At the end of the next section, after some further delving into t, we shall come up with another, tidier, definition.

One further point: what do you do if your found value of t is negative? It is apparent that Eqs 3.1, 3.2 and 3.3 can yield negative values of t, just as easily as positive values. In working through the randomized example, I deliberately did the subtraction of $\bar{y}_1$ and $\bar{y}_2$ so as to get a positive result. However, if the symbols had been allocated differently to the 2 groups of measurements, the found value of t would have had a minus sign (as it did with the paired data). The answer is that for purposes of inserting a found value of t into the t-table, you just ignore any negative sign that it may have. In fact, Eq. 3.1 can be rewritten in the form:

$$t = \frac{(|\bar{y}_1 - \bar{y}_2|)\sqrt{N}}{\sqrt{s_1^2 + s_2^2}} \dots\dots\dots\dots\dots\dots\dots \text{Eq. 3.4}$$

where the vertical bars indicate that the subtraction of the two means is done so as to give a positive result. Eq. 3.2 can be adjusted similarly

Table 3.5   One hundred normally-distributed values arranged in ascending order. They
come from a normal distribution with $\mu = 42.1815$ and $\sigma = 18.477$

Item no.	Value	Item no.	Value	Item no.	Value	Item no.	Value	Item no.	Value
00	0.2	20	26.6	40	37.5	60	46.9	80	57.7
01	2.0	21	27.3	41	38.0	61	47.3	81	58.4
02	4.2	22	27.9	42	38.5	62	47.8	82	59.1
03	7.4	23	28.5	43	38.9	63	48.3	83	59.8
04	9.8	24	29.1	44	39.4	64	48.8	84	60.6
05	11.8	25	29.7	45	39.9	65	49.3	85	61.3
06	13.5	26	30.3	46	40.3	66	49.8	86	62.1
07	14.9	27	30.9	47	40.8	67	50.3	87	63.0
08	16.2	28	31.4	48	41.3	68	50.8	88	63.9
09	17.4	29	32.0	49	41.7	69	51.3	89	64.8
10	18.5	30	32.5	50	42.2	70	51.9	90	65.9
11	19.5	31	33.0	51	42.7	71	52.4	91	67.0
12	20.5	32	33.5	52	43.1	72	52.9	92	68.1
13	21.4	33	34.0	53	43.6	73	53.5	93	69.4
14	22.2	34	34.6	54	44.0	74	54.1	94	70.9
15	23.0	35	35.1	55	44.5	75	54.6	95	72.6
16	23.8	36	35.6	56	45.0	76	55.2	96	74.5
17	24.6	37	36.0	57	45.4	77	55.8	97	76.9
18	25.3	38	36.5	58	45.9	78	56.4	98	80.1
19	26.0	39	37.0	59	46.4	79	57.1	99	82.4

3.4   Further Insights into the *t*-Test

3.4.1   Let us generate our own t-distribution!

To gain some further insights into the *t*-test, we need to know something about
the *t*-distribution on which it is based. This can be approached in 2 ways:

(1) through mathematical theory which would be the elegant and rigorous
way, and

(2) by generating our own *t*-distribution by random sampling from a normal
distribution. This is simple and instructive and is set out in the following 14-step
exercise.

Step 1. First we need a population of normally-distributed measurements.
This is provided in Table 3.5 which lists 100 such measurements conveniently
listed in ascending numerical order from 00 to 99.

Step 2. Consult the table of random numbers (Appendix A1) and, starting
with the first row and taking the digits in pairs as they come, set down the first
4 pairs as row 1: 20 17 42 28.

Step 3. Without any gaps, continue to set down the random numbers as row
2, underneath row 1 to give:

Row 1:	20	17	42	28
Row 2:	23	17	59	66

54

Step 4: From Table 3.5 obtain the *measurements* corresponding to each random number, e.g. random number 20 corresponds to a measurement of 26.6 (units not important). This gives:

Row 1:   26.6   24.6   38.5   31.4
Row 2:   28.5   24.6   46.4   49.8

Step 5: Regard the two rows as a set of 4 vertically-paired experimental measurements and subtract each row 2 value from its corresponding row 1 value, keeping the plus or minus sign correct. These *difference values*, which we shall designate as *h*, are:

h: -1.9 0.0 -7.9 -18.4

Step 6. Using Eqs 1.1 and 1.3 calculate $\bar{h} = -7.05$ and $s_h = 8.2819$.
Step 7. Using Eq. 3.3 calculate *t*:

$$t = \frac{\bar{h}\sqrt{N}}{s_h} = \frac{-7.05\sqrt{4}}{8.2819} = -1.702.$$

Step 8. Go back to Step 2 and continue to read the random numbers from where the last set ended. Repeat all the above steps from 2 to 7 so as to get a second value of *t*.

Step 9. Repeat Steps 2–7 at least 50 times (preferably 100 times), moving steadily through the random number table as you go and thereby collecting a large series of *t*-values.

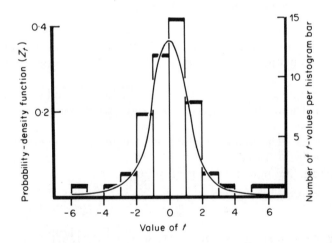

Fig. 3.3   Histogram of 50 values of *t* obtained by random sampling of groups of 4 pairs of measurements from the normal distribution in Table 3.5. Superimposed on the histogram is the probability-density function (Z_t) for the *t*-distribution (with 3 degrees of freedom)

Step 10. Summarize these *t*-values as a histogram. This gives the diagram in Fig. 3.3, which was obtained by the above procedure. It shows a symmetrical distribution, not unlike a normal distribution in shape, with a central value of zero. This histogram is a rough representation of the '*t*-distribution for 3 d.f.'. Superimposed on the histogram is the theoretical *t*-distribution curve for 3 d.f. (next step).

Step 11. We can imagine that if we went on with this process of random sampling 4 pairs of measurements from a normal distribution to build up, say 1 million values of *t*, that we could get a very smooth and symmetrical distribution the outline of which would form a bell-shaped curve. (This whole discussion is very similar to the one we used when explaining the normal distribution curve in Chapter 2.) The *t*-distribution curve is, in fact, very like the normal curve in shape except that it has higher tails. Like the normal distribution, the *t*-distribution can be expressed mathematically, and in Fig. 3.3 the theoretical *t*-curve for 3 degrees of freedom (from Bliss, 1967) has been superimposed on the histogram of the experimental values. Even with only 50 experimental values of *t*, the fit of the histogram to the theoretical curve is quite good.

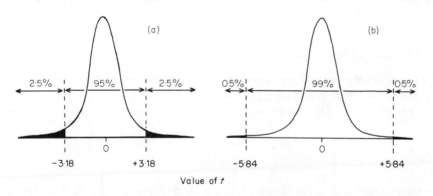

Fig. 3.4   a, the *t*-distribution, for 3 degrees of freedom, with the area under the curve divided so as to yield a broad central zone enclosing 95% and two tails each with 2.5% of the area. The cutoff points for the tails are at $t = \pm 3.18$, the value of *t*-corresponding to a probability of 5%. (b), as (a) but with cutoff points at $t = \pm 5.84$ (corresponding to 1% probability) so as to yield two small tails each with 0.5% of the area under the distribution curve

Step 12. As with other frequency-distribution diagrams, our main interest in the *t*-distribution of Fig. 3.3 is in the *areas* under the curve, because these areas are a measure of the frequency-of-occurrence, or probability-of-occurrence, of particular values of *t* during random sampling. So look now at Fig. 3.4a where the *t*-distribution is divided into 3 zones: a broad central area which encloses 95% of values of *t* and two tails each with 2.5%. In making these divisions one looks up the *t*-table for 3 d.f. and $P = 5\%$ and finds the value $t = 3.182$. This is actually to be read as $t = \pm 3.182$. This tabulated value is the value of t *inside* which 95%, and *outside* which 5%, of t-values obtained by random sampling from a normal distribution can be expected to lie. To repeat this: values of t lying outside of ± 3.182 will only occur, on average, on 5%, or 1 in 20, of random samplings of 4 pairs of observations from a normal distribution.

Step 13. In a manner similar to the above, we can divide up the t-distribution into a still broader, central zone that incorporates 99% (and excludes 1%) of t-values (Fig. 3.4b). The abscissa values for this zone are at ± 5.841, i.e. the tabulated value of t for 3 d.f. and $P = 1\%$.

Step 14. The t-distribution which we see as a single bell-shaped curve in Figs 3.3 and 3.4 is actually a *family* of bell-shaped curves, with a slightly differently shaped curve for each value of the d.f. The latter can range from 1 to infinity. For this reason, the values of t, corresponding to 5% and 1% of the area under the curves, change progressively with the d.f. as we move down through the table. When the d.f. becomes large, say >30, the t-distribution approximates very closely to a normal distribution. At d.f. $= \infty$, the t and the normal distribution are identical and a t-value of ± 1.96 encloses 95% of the area under the distribution, as happens with the normal curve (when we mark off the abscissa in units of standard deviation on either side of a mean value of zero).

3.4.2 Our dilemma: one population or two?

Let us forget about random sampling from a normal distribution and go back to our 8 randomly-obtained pipette deliveries, of which 4 were drainage and 4 were blow-out. The means were:

$$\text{drainage: mean } (\bar{y}_1) = 4.85 \text{ mg.}$$
$$\text{blow-out: mean } (\bar{y}_2) = 26.60 \text{ mg}$$

Now, bearing in mind that a *sample* is to be regarded as representative of an underlying *population*, how are we to interpret the above values of $\bar{y}$? There are 2 possibilities:

(1) $\bar{y}_1$ and $\bar{y}_2$ should be regarded as samples from *two* underlying populations with different mean values μ_1 and μ_2, i.e. a 'drainage' population and a 'blow-out' population, whose means are *not* coincident (Fig. 3.5a).

(2) $\bar{y}_1$ and $\bar{y}_2$ should be regarded as samples from *one* underlying population (Fig. 3.5b), the difference between $\bar{y}_1$ and $\bar{y}_2$ having arisen purely by chance, through 4 low readings falling into one group that we labelled 'drainage' and 4 high readings falling into the other group that we labelled 'blow-out'.

Fig. 3.5 illustrates these two possible interpretations of the data. The purpose of doing the t-test is to establish *which* interpretation is correct — or rather, to see whether we can eliminate one of the possibilities as being so improbable that it can be dismissed. This leads us into that peculiar device, the *null hypothesis.*

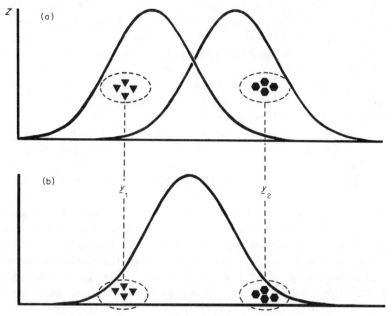

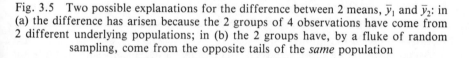

Fig. 3.5 Two possible explanations for the difference between 2 means, $\bar{y}_1$ and $\bar{y}_2$: in (a) the difference has arisen because the 2 groups of 4 observations have come from 2 different underlying populations; in (b) the 2 groups have, by a fluke of random sampling, come from the opposite tails of the *same* population

3.4.3 The null hypothesis: a peculiar but necessary device

At the first encounter, the null hypothesis (NH) strikes many people as a peculiar and back-to-front way of dealing with a problem. But after repeated exposures, the logic of it does sink in and one becomes accustomed to using it. Eventually one accepts it as an orderly way of trying to resolve the dilemma that was posed in the previous sections, namely: Are our two mean values, $\bar{y}_1 = 4.85$ and $\bar{y}_2 = 26.60$, to be regarded as estimates relating to *two* different underlying populations with means μ_1 and μ_2? Or are they estimates of the mean (μ_3) of *a single population*—estimates which were rather far apart because of a fluke in random sampling that happened to select from the lower and upper tails of the *same* distribution?

Let us rephrase the second alternative so that we express it, not as a question, but as a definite assertion, or *hypothesis*: 'The two sample means $\bar{y}_1$ and $\bar{y}_2$ come from a *single* underlying distribution and differ from each other by no more than can reasonably be explained by random sampling fluctuations.'

This is an example of a *null hypothesis*. It is a statement to the effect that there is no, or *a null*, difference in the underlying populations from which the two samples were drawn, and it asserts that the difference between the two means, $\bar{y}_1$ and $\bar{y}_2$, is due *purely* to the chance effects of random sampling.

We have to think of setting up the NH as a conditioned reflex: it is something we set up *automatically* whenever we do a Test of Significance, even if it seems blatantly obvious to the naked eye that the two $\bar{y}$-values could not *possible* come from the same underlying distribution. We are to regard it as part of our orderly approach to resolving the dilemma of *one* population or *two*.

Having set up the NH, we then test it. The test consists in:

(a) calculating the 'found' value of t, as we have already done;

(b) interpolating this found value into the t-table at the appropriate number of d.f.; and

(c) reading out from the t-table the probability value associated with the 'found' value of t.

Now here is the essence of the whole argument: the probability we get from the t-table is the probability of the 'found' t arising as a result of random sampling from a *single* underlying population, i.e. it is the probability of getting that particular result (or values more extreme than it) *if* the NH were true. If this probability is very low, then we reject, or knock down, or discard, the NH: we can no longer give credence to it as a reasonable interpretation of the data. Having thus dismissed the NH, we are left with the alternative possibility: that the two means, $\bar{y}_1$ and $\bar{y}_2$, are samples from *two different* underlying populations or, in the conventional phrase, that they are significantly (or highly significantly, as the case may be) different from each other. Thus, we have now uncovered another definition of the word *significant*, expressed in terms of rejection of a NH.

An additional note about the word 'significant': a statistically significant result is not necessarily 'significant' in the everyday sense of being important or worthwhile. For example, a drug may be shown by a clinical trial to prolong the lives of patients from an average of 16 days to an average of 23 days. This bit of information may be statistically significant in the sense that random-sampling fluctuations cannot reasonably account for the difference in survival times. However, the patients themselves may not think that the extra 7 days are so 'significant', especially if the drug is unpleasant to take.

3.4.4 Why 5% and 1%?

The answer is straightforward: it is purely a matter of convention. There is nothing sacred about 5% and 1% as the boundaries for the different levels of significance. They are not universal physical or mathematical constants, and are merely levels of probability which down the years have been found useful and convenient for interpreting the results of statistical calculations. So we should avoid thinking that there is a quantal jump in significance between a P-value of 4.5% and one of 5.5%, even though the former would be categorized as 'significant' and the latter as 'not significant'. The fact is that if we only have a single experiment giving a P-value in the region of 4–6%, we would certainly want to repeat that experiment a good number of times before coming to *any* very definite conclusion. However, if say, on *each* of 5 independent occasions one mean was consistently higher than the other and the *P*-value was 10% then taking the 5 experiments together, the probability of this occurring through sampling fluctuations is $(1/10)^5$, or 1 chance in 100,000. (Note that this example is rather artificial since you would be very unlikely in 5 repeat experiments to get a *P*-value of exactly 10% each time.) We would therefore judge the difference in the two groups of measurements as highly significant when considered together, even though no individual experiment may have exhibited a significant difference.

Let us return for a moment to the justification for taking our significance boundaries at 5% and 1%. In addition to these figures being nice round numbers (5% = 1 chance in 20), they also tend to coincide with our judgement of significance 'by eye'. Fig. 3.6 shows the drainage/blow-out data progressively modified by reducing $\bar{y}_2$ while keeping $\bar{y}_1$ and the standard deviations constant. Put your hand over the bottom part of the diagram and try to assess at what point in the graded series from *A* to *E* you would cease to regard the two groups as 'significantly different by eye'. Is it somewhere close to *C*? Then your naked-eye assessment agrees with the *t*-test.

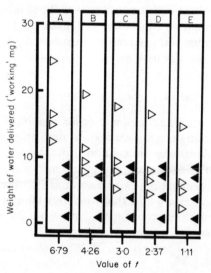

Fig. 3.6   Results of a series of 5 fictitious 'drainage/blow-out' experiments in which the within-group scatter is kept constant but the mean of the 'blow-out' group (open triangles) is progressively lowered, while the 'drainage' group (solid triangles) is kept constant. As judged by the *t*-test, the differences are A and B: highly significant ($P<1\%$); C, significant ($P<5\%$); D and E, not significant ($P>5\%$)

3.4.5   The monkey at the typewriter

In the above sections we show how the *t*-test and the null hypothesis allow us to make a *probability statement* about whether two groups of measurements are 'significantly different'. Notice that we do not try to say *absolutely definitely*, or *categorically*, that there *is* a 'real' difference between the 2 groups. Instead, we content ourselves with the circumlocution of stating that 'since there is such a low probability of the difference arising through random-sampling fluctuations, it is *much more reasonable to conclude* that the difference is real'. However, even if we are rejecting the null hypothesis at the $P < 0.1\%$ level, we should always have at the back of our minds the possibility that the data *could*, by a fluke of random sampling, have arisen from a *single* underlying population and that the observed difference in the means is quite spurious. As scientists, how do we meet this criticism from people who 'want to know the truth'? Mainly we meet it by repeating our experiments 3 or 5, or more, times. So, if in 5 repeats we get each time the same mean higher than the other and a P-value of less than 0.1%, then the chance of this arising 5 times in succession is less than 1 in $(10^3)^5$, or less than 1 in 10^{15}. Even then, we, as cautious scientists, are not saying it could *never* happen. But we should feel reasonably confident in believing that we *have* observed a 'real' difference even though there is a truly minute chance that we might be wrong. When one is dealing with astronomically large numbers like 10^{15}, one is discussing probabilities in the same category as whether a monkey at a typewriter could produce a Shakespeare sonnet. One may be prepared to admit the theoretical possibility of this happening perhaps *once* (given enough monkeys, typewriters and time), but for practical purposes the chance of it happening within a reasonable period of observation is so close to nil as not to be worth discussing.

3.4.6 Can we prove that two groups are the same?

We have seen in the foregoing that the rejection of a NH leads us to conclude that the difference in two means, $\bar{y}_1$ and $\bar{y}_2$, is 'real' and cannot reasonably be ascribed to random-sampling fluctuations. But what about the opposite situation? If we *fail* to reject the NH, does that mean that $\bar{y}_1$ and $\bar{y}_2$ come from the *same* underlying population? The answer is no. We *cannot* draw that conclusion. All we can conclude is that they *may* have come from the same population; but equally, they could have come from 2 different populations whose μ-values were sufficiently close as not to be distinguishable in samples of the size available. *The absence of a significant difference is thus not a proof of sameness*, although for certain everyday purposes we may choose to treat it as such.

Another possibility for error is that by a fluke of random sampling, we have picked, say 4, observations from the upper tail of one distribution and 4 from the lower tail of another, quite separate, distribution, as shown in Fig. 3.7. So although the two underlying distributions have quite distinct mean values, μ_1 and μ_2, an accident of random sampling has given us samples that overlap strongly and give a non-significant t-value. This is the opposite of the case illustrated in Fig. 3.5b where we might reach the false conclusion that our 2 samples came from *different* populations, because a fluke of random sampling

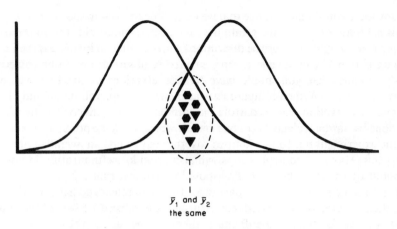

Fig. 3.7 Diagram which shows how the random chance of sampling from the upper portion of one distribution and the lower portion of another distribution can yield samples with identical means, $\bar{y}_1$ and $\bar{y}_2$. Such samples thus fail to reveal that they originated from different underlying populations with means μ_1 and μ_2. This is known as an 'error of the second kind'

gave us groups from the upper and lower tails of the *same* distribution. The only way to reassure oneself that neither of these errors is operating is to repeat the experiment a sufficient number of times with overall consistent results. (See Errors of first and second kind in the Glossary.)

3.5 Other Variations in the *t*-Test

3.5.1 The t-test with more than 2 groups

Although the *t*-test is designed to compare the means of only 2 groups of observations, it can be extended to *multiple comparisons*. But certain precautions and provisos have to be observed, in particular:

(1) If you have, say 6 groups of observations, it would *not* be legitimate to take the groups in any arbitrary combination of pairs and compare all the possible pairs with each other by *t*-testing. This would inevitably lead you into judging as significant, differences which were not.

(2) Likewise, for the same reason it would be illegitimate to pick out the highest and lowest groups and apply the *t*-test just to them.

However, multiple comparisons are allowable under the following conditions:

(1) If Analysis of Variance (Chapter 7) shows that there is significant heterogeneity of group means.

(2) If the comparisons to be made have been identified *in advance of doing the experiment*. For example, you may wish to compare a series of treatment groups with a pre-identified baseline, or control, group to determine which treatments have produced an effect significantly above or below baseline. This kind of multiple comparison can be done by the *t*-test but, if you know this

in advance, you should arrange to have extra observations in the baseline group. This is because the baseline group is to be the reference point for all the other groups and should therefore be determined with extra reliability. The quantitative advice is that if you have k groups, each of N observations, to be compared with baseline, then you should have $N\sqrt{k}$ observations on baseline, e.g. with 4 test groups, each to be compared with a control group, you should have twice as many observations in the control as in each test group. This could conveniently be done by having 2 identical control groups which would be processed separately in the experiment but finally lumped together to provide an overall control for the t-test. Having 2 control groups will also provide useful information on the amount of variation between 2 supposedly identical groups.

(3) In making multiple comparisons, it is recommended that a more strict significance criterion be applied, e.g. requiring a P-value $\leqslant 2.5\%$ for 'significant' and $\leqslant 0.5\%$ for 'highly significant'.

3.5.2 One-tail and 2-tail t-tests

Sometimes we wish to make a comparison of 2 means, like $\bar{y}_1$ and $\bar{y}_2$ from the the drainage/blow-out experiment, where we know beforehand that one of the means ($\bar{y}_2$) *should not be smaller* than the other because of the physical circumstances associated with it, e.g. a set of results obtained by blowing out the last drop from the pipette should not be *smaller* than a set obtained by simple drainage. The blow-out results may not be *significantly* greater, but at least they should not be smaller, unless there is some superimposed bias which unexpectedly counteracts the effect of blowing out the last drop.

This consideration leads to another type of null hypothesis (NH) in which instead of asserting that 'There is no significant *difference* between the 2 means' we say 'the blow-out mean, i.e. $\bar{y}_2$, is not significantly *greater* than the drainage mean $\bar{y}_1$'. In other words, we have to identify one of the means as plausibly being greater, before starting the calculation. We then calculate t in the same way as before but we adopt a different procedure to compare the found value of t with the tabulated values. Effectively, the new version of the NH is equivalent to ignoring one of the tails of the t-distribution and taking a smaller tabulated value of t which therefore has a bigger probability area in its tail.

Suppose, for example, in Sections 3.3.1–3.3.4, the difference between blow-out and drainage was reduced so that the t-value was 1.943, which corresponds to $P = 10\%$ in a 2-tail test (d.f. = 6). This would lead to the conclusion that there was no significant difference between blow-out and drainage. If, however, we decide beforehand that the relevant question was whether blow-out yielded significantly *more than* drainage (as distinct from merely being *different*, without specifying more or less), then we could employ the 1-tail t-test, and a found $t = 1.943$ would correspond to $P = 5\%$ in such a test. This would be interpreted as 'significant (barely)'. Thus, we can see that in special circumstances it may be advantageous to use the 1-tail test.

But in reality the main reason for mentioning the 1-tail t-test is not so much because of its potential utility, which is likely to be slight in the present context, but because the concept of '1-tailed-ness' comes up in discussions of other tests of significance and it is desirable therefore to have some idea of its meaning.

3.6 Comparison of Variances: The F-Test

In discussing the t-test we pointed out that one of the preconditions for doing it was that there should not be a significant difference in the standard deviations of the 2 groups being compared. The test that we use to investigate this matter is the F-test and, before going any further, we need to bring in the term *variance*. Variance is simply the square of the standard deviation; or you can think of standard deviation as being the square root of the variance. The variance is usually written as s^2.

The calculation of F is easier than that of t and we shall now do an F-test on the drainage/blow-out pipette data. Here (Table 3.2) we found the standard deviations to be:

$$\text{drainage:} s_1 = 3.428,$$
$$\text{blow-out:} s_2 = 5.320.$$

At a casual glance these do not look very different, although s_2 is 50% greater than s_1. Is this a significant difference? If not, how big a difference would be needed before it *would* be significant?

3.6.1 F-test: procedure

Step 1. Substitute the values of s into the following formula, arranging for the larger value of s to be on the top line of the fraction:

$$F = \left(\frac{s_2}{s_1} \right)^2 \quad \dots\dots\dots\dots\dots\dots\dots\dots\dots\text{Eq. 3.5}$$

(depending on how the subscripts 1 and 2 have been allocated to the s-values, this fraction might have to be inverted so as to have the larger s on the top line). This gives:

$$F = \left(\frac{5.320}{3.428} \right)^2$$
$$= 1.5519^2$$
$$= 2.41$$

This is our *found value* of F.

Step 2. Compare this found value of F with the values of F given in the F-tables (Appendix A3). As with the t-table, we first have to determine the number of degrees of freedom (d.f.). With F, let us suppose we have n_1 observations giving rise to s_1, and n_2 observations giving rise to s_2 (we are taking the general case where n_1 may not be the same as n_2). Then the degrees of freedom are $(n-1)$ which in this example are 3 and 3, usually written as 'd.f. $=3,3$'.

The F-tables have to be entered with the d.f. of the *larger* variance determining the vertical column, and the d.f. of the *smaller* variance the horizontal row in the table. (Note that variance is also known as 'mean square'.) Each table then provides, at the point of intersection, the *tabulated value* of F, one table being for $P=5\%$ and the other for $P=1\%$. Thus, for d.f. $=3,3$ we find in Appendices A3A and A3B, respectively:

$$F= \;\;9.28 \text{ for } P=5\%$$
$$F=29.46 \text{ for } P=1\%$$

We therefore interpret our *found* value of $F=2.41$ as 'not reaching the $P=5\%$ level and thus *not* significant'. We can go on to say that for 2 sets of 4 measurements to have standard deviations that are barely significantly different (i.e. just reaching the $P=5\%$ tabulated value of F) the ratio of variances has to be $9.28:1$, which means that the ratio of standard deviations would be $3.05:1$. In our drainage/blow-out example the standard deviations are *not* significantly different and we were therefore (on this criterion at least) not disbarred from doing the t-test.

3.6.2 F-test explained

The theoretical background to the F-test is very similar to that for the t-test. In both, we are dealing with

(a) measurement data;

(b) random samples;

(c) samples from underlying normal, or
approximately normal, distributions;

(d) a comparison of 2 samples with each other.

The difference is that in the t-test we are interested in comparing the *means* of 2 samples and in the F-test with comparing their *variances*.

Implicit in doing an F-test is the setting up of a NH which states that: 'The difference in variances of the two samples is no greater than can reasonably be explained by random sampling fluctuations from a *single* underlying distribution (or from 2 distributions which may have different values of μ but have the same value of σ)'.

The test of the NH is whether the *found* value of F can 'reasonably be explained by random-sampling fluctuations'.

Like the t-distribution, the F-distribution exists as a family of smooth, humped curves, except that the F-curves are asymmetric. And like t, one can produce

Table 3.6 Statistical tests for the analysis of normally-distributed grouped data

Test	Purpose	Restrictions[a]	Notes	Ref.
Rankit plot	To check normality of small-group data	Need at least about 5 measurements per group	Also facilitates comparisons of mean and S.D. if several groups are being assessed.	Chapter 4
Probit plot	To check normality of a large group of data	Need ~25 replicate measurements at least	Also estimates mean and S.D.	Chapter 2
Student t-test	Comparison of the means of 2 groups of data	Data should be from a N.D.; variances of the 2 groups not significantly different	Can be extended to >2 groups; Different formula for paired data	Chapter 3
Behrens–Fisher d-test	Comparison of the means of 2 groups of data but where the t-test cannot be used because the variances are significantly different	Data should be from a N.D. but variances may be significantly different	d-table not given in this book	See Bliss (1967, p. 216)
F-test	Comparison of the variances of 2 groups of data	Data should be from a N.D., or approximately N.D.	Extensively used in analysis of variance	Chapter 3
Bartlett test	Comparison of variances of >2 groups of data	Test fairly sensitive to departures from N.D.	Uses the χ^2-statistic	Chapter 7
Analysis of variance	Determining which factors in an experiment have influenced the results	As for F-test	Uses the F-statistic	Chapter 7

[a]N.D. = normal distribution.

an approximation to the F-distribution by random sampling from a normal distribution, as we did for t in Section 3.4.1.

Also like t, the *tabulated* values of F reflect the relative areas under the F-distribution curve, and the associated P-values represent the relative probability-of-occurrence of particular F-values in a long run of random sampling from a normal distribution. So, if we did experimental sampling of 2 groups of 4 observations from the normal distribution in Table 3.4, then on an average of 5% of occasions we would expect to get an F-value as high as, or higher than, 9.28. On the other 95% of occasions the long-run values of F would be less than 9.28.

In analysing the results of a single $4 + 4$ observations experiment, an F-value of 9.28 could mean one of two things:

(1) the 2 groups of observations reflect samples from 2 different underlying populations with different standard deviations, σ_1 and σ_2, or

(2) we have encountered the 1 in 20 chance event which has given a high value of F from a *single* underlying population, because we have selected some highly scattered measurements in one group (and therefore a large s) and tightly bunched observations (and therefore a small s) in the other.

Following the usual convention, we describe such a single result as 'significant', but not knowing whether explanation (1) or (2) represents the 'truth'. If, however, we go on to repeat the experiment several times, we should be able to decide which is the more likely explanation.

3.6.3 Comparing more than 2 groups

The F-test, as described above, is satisfactory for comparing 2 variances. Often, however, we have more than 2 groups of data and we may wish to find out whether the variances are homogeneous for the different groups, or whether they exhibit significant *heterogeneity*. For this purpose Bartlett's test may be used, as described in Chapter 7.

3.7 Summary of Statistical Tests for Analysis of Normally-Distributed, Grouped Data

For convenience of reference, Table 3.6 is appended to summarize the various tests which may be applied to normally-distributed grouped data.

4

More about Differences

The statistician must be treated less like a conjurer whose business is to exceed expectation, than as a chemist who undertakes to assay how much of value the material submitted to him contains

R. A. Fisher

4.1 Staying Normal or Going Non-Parametric

In the previous chapter we discussed two Tests of Significance — the t-test and the F-test — which are only applicable to samples from normal, or approximately normal, distributions. A common dilemma, especially in the early stages of a biological investigation, is not knowing whether the measurements being gathered should be treated as normal or not. (Note that the present discussion is restricted to *measurement* data: proportions and counts are considered in Chapters 5 and 6.) Occasionally there may be little problem if there is a large mass of measurements, as in Chapter 2, where 135 weight readings were shown to fit a normal distribution reasonably well. But in research, this is unusual. Instead, there are much more likely to be small-sized groups of data to be analysed for possible differences. The question arises as to how this can be done without knowledge of the underlying distribution. Conversely, how can the nature of the underlying distribution be uncovered if the data are gathered in small groups? Let us now see what can be done about this chicken-and-egg situation by means of one of the 3 approaches labelled as *precedence, prudence* and *probing*.

4.1.1 Precedence

Unless the investigation is in a completely unexplored subject, the statistical methods used by recent workers should provide a starting point. But the fact that there is precedence for a particular statistical treatment does not mean that that treatment is the best possible, or is even satisfactory. Scientific journals vary greatly in the level of statistical expertise expected from contributors, so the methods established by precedence should be taken only as an initial guide. Nimmo and Atkins (1979) pointed out that in the issues of the *Biochemical Journal* for early 1979 over one-half of the papers included some form of statistical analysis. The overwhelming majority of these placed reliance on the traditional parametric statistical methods such as Student's t-test, despite the serious flaw of not knowing whether the underlying distribution was normal or not.

4.1.2 Prudence

Some would say that in the initial stages of an investigation the prudent approach to statistical analysis of groups of data would be not to make *any* assumptions about the nature of the underlying distributions. This would then avoid such errors as applying the t-test to non-normal data. Tests which avoid assumptions about the normality of the underlying distributions are the so-called *non-parametric* tests dealt with later in this chapter. At first sight they seem very attractive. However, like insurance policies, when you 'read the fine print' you find there are some drawbacks and, in particular, the non-parametric tests:

    (1) Tend to be somewhat less sensitive than the corresponding parametric

procedures like the t-test. This means that to detect significant differences between groups of data you need more observations in each group, and also the average differences themselves have to be greater.

(2) Although applicable to samples from both normal and non-normal distributions, most of the tests assume that the different groups of data in an experiment come from distributions of *the same shape*, i.e. the shape may be unknown, but is assumed to be constant. The different groups thus may vary only in the positions of their medians. The assumption of constancy of shape of the underlying distribution is difficult to prove, so that by 'going non-parametric' you may not quite reach the intellectually pure state of freedom from unverifiable assumptions. Mainly what you have done is to avoid the assumption of normality. On the credit side, however, it must be said:

(3) that the non-parametric tests tend to involve less calculation than the parametric ones, and

(4) with certain types of data, such as survival times in animal-infection experiments, non-parametric tests may be the only ones available.

4.1.3 Probing

By this I mean that at a very early stage in an investigation where measurement data are being gathered, a deliberate effort should be made to probe the nature of the underlying distribution. In particular, it is highly desirable to know whether the data can reasonably be regarded as samples from underlying normal distributions; or whether they should be 'normalized', e.g. by taking logs; or whether they are definitely not normal and should therefore be analysed by the non-parametric methods. There are considerable advantages in having normal, or normalized, data. In particular, the technique of Analysis of Variance (Chapter 7) is greatly facilitated, and Biological Assays (Chapter 10) are easier to evaluate. So it is worthwhile to put in the effort to categorize the data as (a) normal or (b) normalizable or (c) not normal or normalizable. A useful rough check on normality of small groups of data is provided by the rankit procedure which comes next.

4.2 Rankits to the Rescue

Chapter 2 showed how with a large assemblage of measurement data — say, in the order of 100 replicate observations — it is possible to make an objective judgement of the normality of the data with the aid of probits. The quantities known as *rankits* serve a similar function, except you do not need so many observations to use them. Groups of as few as 5 replicate observations can give some information about normality, while groups containing 10–15 replicates may be fairly definitive.

4.2.1 Introduction to rankits

The use of rankits is now illustrated by applying them to the 9 pipetting results from Chapter 1. Expressed in 'working mg', these were:

$$\text{Pipette A:} \quad 37.5, \quad 16.8, \quad 22.7$$
$$\text{Pipette B:} \quad 11.6, \quad 24.7, \quad 12.0$$
$$\text{Pipette C:} \quad 24.8, \quad 27.7, \quad 5.9$$

The procedure for using rankits is as follows.

Step 1. Arrange the observations in *rank order*, starting with the highest value as no. 1 and the lowest as no. 9. This may be done as:

Rank no.	1	2	3	4	5	6	7	8	9
Value (working mg)	37.5	27.7	24.8	24.7	22.7	16.8	12.0	11.6	5.9
Pipette	A	C	C	B	A	A	B	B	C

Step 2. Look up the rankit table (Appendix A5) and extract the rankit values for a group size of $N = 9$ observations. This gives the following set of figures:

Rank no.	1	2	3	4	5	6	7	8	9
Rankit	1.485	0.932	0.572	0.275	0	−0.275	−0.572	−0.932	−1.485

Note that the rankits are symmetrical about the middle value.

Step 3. Plot a graph of 'working mg' against rankit (Fig. 4.1). Join up the points to give a stretched-out zig-zag line.

Step 4. Calculate the mean and standard deviation of the 9 observations. From Chapter 1 we have:

$$\bar{y} = 20.41$$
$$s = 9.75$$

Step 5. Calculate $\bar{y} - s = 20.41 - 9.75 = 10.66$ and $\bar{y} + s = 30.16$. Superimpose on the zig-zag rankit plot a straight line by joining the points at (wt = 10.66, rankit = −1.0) and (wt = 30.16, rankit = +1.0).

Step 6. Interpretation: If the 9 experimental observations are a random sample from an underlying normal distribution, then the *observed rankit* line (solid) should zig-zag in a random fashion fairly closely around the fitted theoretical rankit line (dashed). The results in Fig. 4.1 meet this requirement as well as can be expected. However, you should note that a rankit plot on a single set of 9 observations is unlikely to be sufficient for a definite anwer — but at least it may reveal a gross *departure* from normality. Nevertheless, if you are doing several runs of an experiment, the routine use of the rankit plot can give either worthwhile assurance of normality or indicate that the data are non-normal. Thus, non-normal data give a rankit plot in which the experimental rankit line

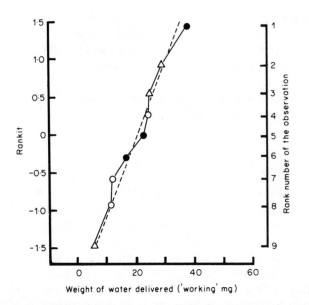

Fig. 4.1   Rankit plot of the 9 pipetting results from Chapter 1. Each pipette is represented by a different type of point (●: a, ○: b, △: c). The right-hand ordinate scale is for convenience in rankit plotting when several groups of 9 observations have to be processed

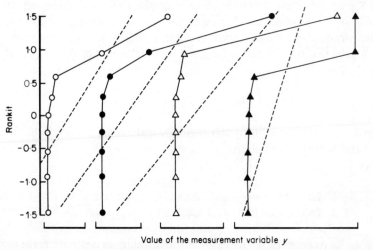

Fig. 4.2   Four sets of 9 observations from a log-normal distribution subjected to rankit plotting before 'normalization'

(a) departs in a *consistent* fashion from the independently fitted straight line, and

(b) shows less criss-crossing of the fitted straight line than would be expected with normal data.

For example, Fig. 4.2 shows the rankit plots of 4 sets of 9 observations each where

(a) the dotted straight lines are obviously poor fits to the experimental points

(b) the rankit plots do not oscillate randomly around the dotted lines

(c) the rankit plots all have more or less the same shape—like an inverted letter L and are not simple zig-zags around a straight line.

In fact, the 4 sets of observations in Fig. 4.2 come from a *log-normal* distribution, and we shall soon see how, by 'normalizing' the data, they then give rankit plots resembling Fig. 4.1.

4.2.2 Rankit plots from genuine normal data

The question we may now ask is *by how much* the experimental rankit plot can deviate from the fitted straight line and still be within the *limits of random sampling fluctuations of a normal distribution*. Perhaps the most direct way to answer this is to do some rankit plots on samples selected randomly from an underlying distribution which we know to be genuinely normal. This also provides a worthwhile exercise in random sampling from a normal distribution and in plotting rankits. It will also help to illustrate how much variation in the mean and standard deviation there can be in random samples of $N=9$ from *the same* underlying normal distribution. The procedure is as follows:

Step 1. Consult the random number table (Appendix A1) and extract the first set of 9 pairs of digits. (Note in these examples that it is convenient to start at row 1 in the random number table for ease of reference, but the normal practice would be to start at some randomly chosen point in the table.) The result for the first set of 9 pairs of digits (note that we do not skip any, or discard duplicates) is:

$$20 \quad 17 \quad 42 \quad 28 \quad 23 \quad 17 \quad 59 \quad 66 \quad 38$$

Step 2. Consult the table of 100 normally-distributed values (Table 3.5) and extract the normally-distributed value corresponding to each 2-digit random number:

$$20 \quad 17 \quad 42 \quad 28 \quad 23 \quad 17 \quad 59 \quad 66 \quad 38$$
$$26.6 \quad 24.6 \quad 38.5 \quad 31.4 \quad 28.5 \quad 24.6 \quad 46.4 \quad 49.8 \quad 36.5$$

Step 3. Rearrange the normally-distributed values in descending rank order, starting with the highest. To save recopying, it is simplest just to insert the rank numbers from 1 to 9 directly under the list already drawn, e.g. the second last one is no. 1 (49.8), the highest; and the fourth last one is no. 9 (24.6), the lowest (and equal also to no. 8). This gives:

Random no.	20	17	42	28	23	17	59	66	38
Value	26.6	24.6	38.5	31.4	28.5	24.6	46.4	49.8	36.5
Rank no.	7	8	3	5	6	9	2	1	4

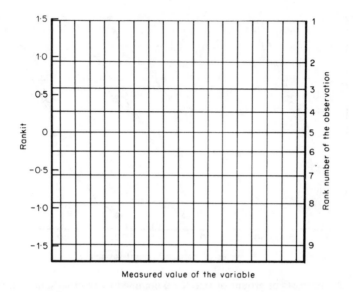

Fig. 4.3 Conveniently ruled graph paper for rankit plots of groups of 9 observations. The subdivision and labelling of the abscissa will necessarily vary from one experimental system to another, but the ordinate labels will be constant so long as $N=9$

Step 4. Look up the rankit table (Appendix A5) for a 9-observation group and plot a graph of y-values against rankit as was done for Fig. 4.1. It is convenient in making these plots to lay out the graph paper as in Fig. 4.3, where the ordinate is labelled with rankits on the left and rankit number on the right, and with horizontal lines ruled across at each rank number. Having made a master sheet, you can then either run off photocopies, or plot the graphs on superimposed tracing paper. A different set of rulings would be needed for groups of other than $N=9$ observations.

Step 5. Calculate the mean $(\bar{y})$ and standard deviation (s) of the 9 observations. This gives $\bar{y}=34.1$ and $s=9.34$.

Step 6. Calculate $\bar{y}-s=24.76$ and $\bar{y}+s=43.44$. Insert the straight line by joining the points at $(y=24.76, \text{rankit}=-1.0)$ and $(y=43.44, \text{rankit}=+1.0)$. This gives rankit plot no. 1 in Fig. 4.4.

Step 7. Go back to Step 1, take a new set of random numbers, continuing where the last set stopped and proceed through to Step 6. Continue in a like fashion to produce two more graphs. Altogether this yields the 4 rankit plots of Fig. 4.4. These provide a good indication of how much irregularity may be expected in rankit plots of sets of $N=9$ randomly-chosen observations from a known and genuine normal distribution. Note that the kinks and bulges in the 4 rankit lines occur quite irregularly, as do the points of criss-crossing, and that both the first and last point in each plot is, on average, no further from the fitted straight line than any of the other points. These are all features to look for when interpreting a set of rankit plots made from experimental data.

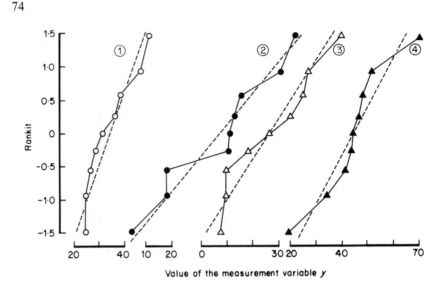

Fig. 4.4   Rankit plots of groups of size $N = 9$ obtained by random-sampling from the normal distribution of Table 3.5. The 4 groups depicted were obtained from the first $4 \times 9 = 36$ pairs of random digits, starting with line 1 of the random number table. Note that the abscissa is split up to facilitate comparison of the shapes of the 4 plots

4.2.3   A temporary diversion from rankits

As a by-product of this exercise it is instructive to inspect the amount of scatter in the $\bar{y}$- and s-values of the four sets of 9 replicates, which, it must be emphasized, were drawn from the *same* underlying normal distribution by random sampling:

Set	$\bar{y}$	s
1	34.1	9.34
2	38.8	21.10
3	26.3	16.42
4	44.5	13.78

Note that $\bar{y}$ varies from 26.3 to 44.5 and s from 9.34 to 21.10. Also bear in mind that these values of $\bar{y}$ and s are to be regarded as estimates of the underlying population parameters $\mu = 42.18$ and $\sigma = 18.48$. However, the difference between this exercise and 'real life' experimentation is that here we *know* the true values of the population parameters because we are dealing with the artificially-constructed, mathematically perfect, normal distribution of Table 3.5.

If you round off this diversionary exercise by calculating the 95% confidence limits (Section 2.7.2) using

$$95\% \ \text{CL} = \bar{y} \pm t.s./\sqrt{N} \dots\dots\dots\dots\dots\dots\text{Eq. 2.5}$$

you will get (taking $t = 2.306$ for $P = 0.05$ and 8 d.f.):

Set	95% CL	Set	95% CL
1	26.9, 41.3	3	13.7, 38.9
2	22.6, 55.0	4	33.9, 55.1

Note that only sets 2 and 4 have confidence limits which include the population mean $\mu = 42.18$. This is a fluke of random sampling since normally we could expect that only 1 set in every 20 would have confidence limits which did *not* include the population mean. If the random number table is entered at the beginning of row 6 instead of row 1, and 4 sets of 9 2-digit numbers extracted and converted into normally distributed measurements by use of Table 3.5, then all 4 sets will be found to give confidence limits which enclose μ.

4.2.4 Rankit plots from non-normal data

You have already seen in Fig. 4.2 the rankit plots on log-normal data and how they contrast with the rankit plots on normal data in Fig. 4.4. What may now be revealed is that both figures use the *same* measurement values randomly chosen from our model normal distribution. So, by comparing the corresponding groups we can seen the beneficial effect of making a logarithmic transformation of the Fig. 4.2 results to get those in Fig. 4.4. The actual method of generating

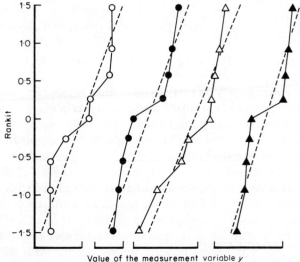

Fig. 4.5 A series of rankit plots from samples exhibiting negative kurtosis. Each of these was produced by taking a set of 20 2-digit random numbers, arranging them in rank order, discarding the 4 highest, 4 lowest and 3 central values, and then treating the 9 that were left as a sample of 9 original observations for rankit plotting. Note that all 4 plots have a 'chair' shape, with the seat facing to the left

the measurements for Fig. 4.2 was to take the *antilog* of each Fig. 4.4 value after dividing it by 10 to make it more manageable arithmetically. This is the reverse of what would happen in real life where we might gather some measurements which gave inverted L-shaped rankit plots initially, but which could then be normalized by converting the measurements to logarithms. Here, for the artificial purposes of this particular exercise, the reverse procedure was carried out.

The type of departure from normality exhibited by a log-normal distribution is *skewness*. Another type of departure from normality is *kurtosis*, where the distribution is symmetrical, but is either unduly peaked and lacking in mid-side values or is unduly flattened at the top and lacking in tail values. These types of departure from normality are much harder to detect by a rankit plot on data where N is as small as 9. They also are harder to detect convincingly by the probit method where you may have $N > 100$. However, we can easily generate artificially a few sets of samples with kurtosis to see what they look like. This has been done in Fig. 4.5 where each rankit plot of $N = 9$ observations was obtained by the following steps.

Step 1. Record the first 20 2-digit random numbers from line 1 of the random number table.

Step 2. Arrange them in rank order from no. 1 (the highest) to no. 20 (the lowest).

Step 3. Impose fairly severe negative kurtosis by chopping off the tails and discarding some of the centre of the distribution. This is done by throwing away observations 1, 2, 3, 4, 9, 10, 11, 17, 18, 19, 20, i.e. more than half of them.

Step 4. Re-number what is left in rank order from 1 to 9, insert the measurement values from Table 3.5 and do an ordinary 9-sample rankit plot (Fig. 4.5).

Step 5. Repeat Steps 1–4 using the 2nd, 3rd and 4th lines of the random number table.

Step 6. Interpretation: Note that although there is inevitably, as you would expect, a strong element of irregularity, all 4 diagrams in Fig. 4.5 have a partial 'chair' shape, with the seats pointing to the left. Any one of these diagrams taken in isolation might well be able to 'pass' as a random sample from a genuine normal distribution. However, the 4 taken together show a clear indication of a systematic and consistent departure from normality. Fig. 4.5 illustrates the value of having several rankit diagrams before reaching conclusions.

4.2.5 Interpenetrating rankits

In random samples from a given normal distribution, the rankit diagrams of each sample may be expected to show some degree of interpenetration, or winding around each other. This is best seen with a group size larger than $N = 9$, and Fig. 4.6 shows what we get with 2 random samples of $N = 20$ observations drawn from the normal data in Table 3.5. Note how the lines intertwine.

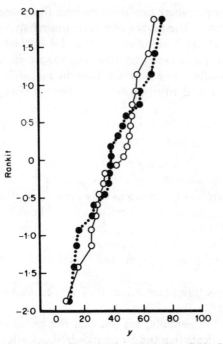

Fig. 4.6   Interpenetrating rankits. Two random samples of $N = 20$ observations from the normal distribution of Table 3.5

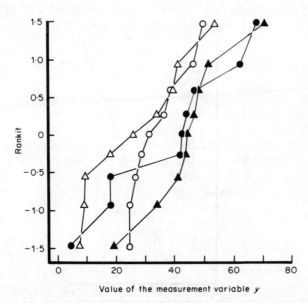

Fig. 4.7   Rankit plots of the 4 sets of $N = 9$ observations from the normal distribution of Table 3.5. This diagram differs from Fig. 4.4 only in not having the abscissa split up and the rankit diagrams separately laterally

78

With smaller groups, where random-sampling fluctuations are liable to lead to greater irregularities, the degree of intertwining may be less apparent. For example, the 4 9-sample rankit plots of Fig. 4.4—which in that figure were deliberately spaced out laterally to make the shapes easier to compare—are plotted together on the same abscissa scale in Fig. 4.7. Here, although the diagrams criss-cross and overlap, none of them is long enough for much intertwining.

4.2.6 Rankits summarized

(1) Rankit diagrams *can never prove normality*, but they may alert you to the existence of non-normality in groups of data.

(2) Done routinely, rankit diagrams may provide worthwhile reassurance on whether grouped measurement data can reasonably be subjected to the statistical methods suitable for random samples from normal distributions, or whether the assumption of normality is suspect.

(3) In a multi-group experiment, the rankit plots can provide additional information:

(a) if convincingly interpenetrating, they suggest that there is no difference in group means;

(b) if separated horizontally, suggest that the group means are different;

(c) if parallel, indicate that the group standard deviations are similar to each other;

(d) if of different slopes, suggest heterogeneity of standard deviations.

(4) Fig. 4.8 shows how a rankit value of 0 corresponds to the arithmetic mean, while values of -1 and $+1$ correspond respectively to $\bar{y}-s$ and $\bar{y}+s$.

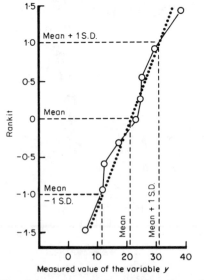

Fig. 4.8   The relationship of the values on the rankit scale to the mean and standard deviation of the group of measurements

4.3 Introduction to Non-Parametric Tests

Since rankits are very much a part of normal-distribution statistics, it might have been more logical to include them in Chapters 1, 2 or 3 which deal with normally-distributed data. However, I felt that these chapters already contained enough information, and that rankits could also be used as the bridge to the non-parametric tests. That is to say, we imagine that the investigator is pursuing a line of work where, either because of rankit plots or from other considerations, it would be inappropriate to treat the data as normal or normalizable. Hence the need to 'go non-parametric'. In fact, the non-parametric tests can also be applied to normal data, but in general we would not use them for this purpose in a well-defined experimental system. Nevertheless, for the sake of convenience and continuity, let us now take some of our already-generated measurement data and analyse them non-parametrically just to see how they turn out. We can justify this by saying that we are adopting the 'prudence' approach of Section 4.1.2. So let us go back to the blow-out/drainage pipetting data which were analysed by the t-test in Chapter 3 and simply reanalyse them by the corresponding non-parametric, Mann–Whitney U-test.

As a second example we shall take the survival times of mice in a bacterial-challenge experiment where the nature of the data precludes t-testing.

4.4 Mann–Whitney U-test

4.4.1 Mechanics of the test

Step 1. Take the 'working mg' data of Table 3.2 and arrange them in ascending order of their numerical values within each group (this is different to rankits where we used a descending order). This gives us:

Drainage (Group 1)	0.6	3.6	7.1	8.1
Blow-out (Group 2)	21.9	24.5	25.8	34.2

Step 2. Assign to each observation its *rank* in the *joint ordering* of the 2 groups, i.e. 0.6 becomes rank no. 1; the value 3.6 becomes rank no. 2, and so on. This gives us the following table of ranks:

Drainage (Group 1)	1	2	3	4
Blow-out (Group 2)	5	6	7	8

Step 3. Obtain the sum of the ranks (R) for each group:

$$R_1 = 1 + 2 + 3 + 4 = 10$$
$$R_2 = 5 + 6 + 7 + 8 = 26$$

Step 4. Check that the arithmetic in Steps 2 and 3 has been done correctly by seeing that:

$$R_1 + R_2 = \frac{1}{2}(N_1 + N_2)(N_1 + N_2 + 1) \dots\dots\dots\dots\dots \text{Eq. 4.1}$$
where $N_1 = N_2 = 4$, ie.
$$10 + 26 = \frac{1}{2}(4 + 4)(4 + 4 + 1)$$
$$36 = \frac{1}{2} \times 8 \times 9$$

which checks.

Step 5. Calculate the U-statistics. First U_1:

$$U_1 = N_1 N_2 + \frac{1}{2}N_2(N_2 + 1) - R_2 \dots\dots\dots\dots\dots \text{Eq. 4.2}$$
$$= 4 \times 4 + \frac{1}{2} \cdot 4 \times 5 - 26$$
$$= 16 + 10 - 26$$
$$= 0$$

Step 6. Calculate the similar U_2:

$$U_2 = N_2 N_1 + \frac{1}{2}N_1(N_1 + 1) - R_1 \dots\dots\dots\dots\dots \text{Eq. 4.3}$$
$$= 4 \times 4 + \frac{1}{2} \cdot 4 \times 5 - 10$$
$$= 16 + 10 - 10$$
$$= 16$$

Step 7. Take the *smaller* of U_1 and U_2 as the *found value* of U. Result: found $U = 0$

Step 8. Arithmetic check that

$$U_1 + U_2 = N_1 \times N_2 \dots\dots\dots\dots\dots\dots\dots \text{Eq. 4.4}$$
$$16 + 0 = 4 \times 4,$$

which checks

Step 9. Consult the U-table for the Tabulated Value of U (Appendix A12) for a $N_1 = N_2 = 4,4$ comparison; the tabulated 5% point of U is 0.

Step 10. Set up a null hypothesis (NH) which states that: 'The difference in medians of the two groups is no greater than can reasonably be explained by random sampling fluctuations.'

Step 11. Accept or reject the NH by the following criteria.

(1) If the found value of U (i.e. the *lower* of U_1 and U_2) is *greater* than the tabulated value, then the NH *cannot* be rejected and the difference between the two groups is *not* significant at the $P = 5\%$ level.

(2) Conversely, if the found U is *less than or equal* to the tabulated, the NH is rejected and the difference is judged to be *significant*. Result: found $U = 0 =$ tabulated U.

∴ NH rejected.

∴ difference is (just) significant.

(Note that the criteria for rejecting the null hypothesis in the U-test are the inverse of those for the t-test where a found value which is *greater* than the tabulated value leads to rejection.)

4.4.2 *Comparison of* U-*test and* t-*test*

A convenient way to compare the 'sensitivities' of the U- and t-tests is to apply them to the sets of fictitious blow-out/drainage measurements which have

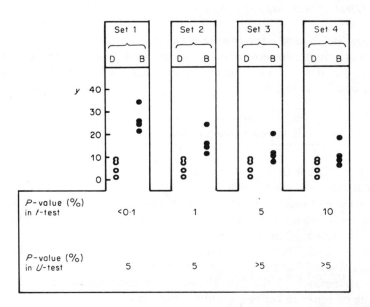

Fig. 4.9 A series of fictitious blow-out/delivery results where the between-group differences have been progressively reduced, and showing the results of analysis by *t*-test and *U*-test

already been used in connection with *t*. This has been done in Fig. 4.9 which presents the original data (Set 1) plus 3 'fictitious' sets. In Set 2, all of the *B*-values have had 10.02 subtracted which shifts them down the diagram and reduces the difference in the group means $(\bar{y}_1 - \bar{y}_2)$ to 11.73. This set has a difference which is significant at the 1% level in the *t*-test but only at the 5% level in the *U*-test. Note that in the *U*-test with 2 groups of 4 observations, provided that the 2 groups do not overlap, the level of significance does not change from 5% no matter how far apart the medians are. By contrast, in the *t*-test of Set 1 the significance level is at $P < 0.1\%$. Set 3 is the critical one. The two groups have been adjusted so as to give a *t*-value of 2.447 which is exactly at the tabulated 5% *t*-value. This was done by subtracting 14.01 from each original *B*-value. Note that as soon as the groups overlap, as in Set 3, the *U*-test ceases to indicate a significant difference. This is not necessarily true with groups having N greater than 4, but does illustrate how, with small groups, the *t*-test can judge as significant a difference which the *U*-test does not. On the other hand, the loss in sensitivity of the *U*-test is counterbalanced by not requiring an assumption either of normality or equality of variances. The *U*-test *does*, however, have the requirement that if the NH is rejected and we wish to conclude that the two samples come from populations with different medians, then we must *assume* that the populations have the same shape. To avoid even this assumption, turn to the Kolmogorov–Smirnov test in Section 4.6.

4.4.3 Dealing with ties

The term 'tie' is used when there are 2 or more identical measurements in a ranking operation. To show how they are dealt with, let us take the data from the blow-out/drainage experiment and alter slightly some of the values so as to give one 2-way and one 3-way tie:

Drainage (Group 1)	0.6	3.6	7.1	7.1
Blow-out (Group 2)	24.5	24.5	24.5	34.2

The procedure with ties is to assign to each tied value the *mean rank* of the values at each position of tying. Thus, the 2 7.1-values occupy rank positions 3 and 4. Each of them is therefore given the mean rank of $3\frac{1}{2}$. (Note that this is different to what was done with rankits where the ties were simply plotted as separate values.) The next rank after the two $3\frac{1}{2}$ entries is no. 5, which has the value of 24.5. This, however, is repeated twice more, so that we have rank positions 5, 6 and 7 all tied. The mean of these 3 is rank no. 6, which is therefore assigned identically to the first 3 readings of the blow-out series. The joint ordering of ranks in the two groups is therefore:

Drainage (Group 1)	1	2	$3\frac{1}{2}$	$3\frac{1}{2}$
Blow-out (Group 2)	6	6	6	8

These figures give the same R- and U-values as the original data, i.e.

Step 3. $R_1 = 1 + 2 + 3\frac{1}{2} + 3\frac{1}{2} = 10$

$R_2 = 6 + 6 + 6 + 8 = 26$

Step 4. Arithmetic check by Eq. 4.1:

$10 + 26 = \frac{1}{2}(4 + 4)(4 + 4 + 1)$

$36 = \frac{1}{2} \times 8 \times 9$

Steps 5, 6. $U_1 = 0$ and $U_2 = 16$, as before.

And the difference between the two groups is still significant, as before.

4.4.4 Another example of tying

Let us get away from pipetting and consider a set of results of a type that occurs very commonly in infection or toxicity tests on animals. Table 4.1(a) shows the results of a mouse-infection experiment in which 12 mice in group A and 10 in group B all received the same challenge dose of bacteria and were then observed daily for death or survival. By visual inspection of the data it can be seen that the Group 1 animals were somewhat more resistant than Group 2, because only 7 out of 12 were killed, as compared with 9/10 in Group 2 (is this difference significant? See Chapter 5). Also, those Group 1 animals that died tended to die later than Group 2, the median death times being about day 5 for Group 1 and day 3 for Group 2. The question to be answered here is: 'Are the median

Table 4.1

(a) *Results of an infection experiment in mice*

Mouse group	Initial no in group	Day of death (post-infection) of individual animals
1	12	2, 3, 3, 4, 4, 5, 5, S, S, S, S, S*
2	10	1, 1, 2, 2, 3, 3, 3, 4, 5, S*

S* = animal survival for 14 days, at which time the experiment was stopped.

(b) *Joint ordering of ranks in the two groups*

Mouse group	N	Ranks of death days	Sum of ranks (R)
1	12	4, 8, 8, 12, 12, 15, 15, 19½, 19½, 19½, 19½, 19½	171½
2	10	1½, 1½, 4, 4, 8, 8, 8, 12, 15, 19½	81½

death-times in Groups 1 and 2 significantly different?' Note that the experiment was terminated on day 14, by which time those animals that were going to succumb had already done so (based on previous experience), and therefore for 7 of the mice we only have a record of S = survival. Since we cannot attach an actual number to the death-day of these animals, we cannot calculate a *mean* survival (or death) time, but can only record the *median*, i.e. the day by which half the animals had died. The *t*-test cannot be applied to such data because S cannot be translated into a number. Note also that although death has been recorded only to the nearest day, it would theoretically be possible to determine the time of death more accurately by keeping the animls under continuous surveillance. This would almost certainly eliminate some of the ties we now have to deal with. But on the other hand the stress of continuous surveillance might have altered the response of the animals to the infection.

To proceed to the task of ranking: as with the examples in Section 4.2.1, we first have to produce a joint ordering of the ranks in the 2 groups. This has been done in Table 4.1(b) which takes account of tied values, as in the previous example. After this ranking operation, we can now proceed with the rest of the calculations which will be given with the same step numbering as in Section 4.4.1.

Step 3. Sums of ranks:
$$R_1 = 171½$$
$$R_2 = 81½$$

Step 4. Arithmetic check: this is very important if the ranking has been complicated, as here:

$$171\tfrac{1}{2} + 81\tfrac{1}{2} = \tfrac{1}{2}(12 + 10)(12 + 10 + 1) \quad \ldots\ldots\ldots\ldots\ldots\ldots\text{Eq. 4.1}$$
$$253 = \tfrac{1}{2} \times 22 \times 23$$
$$= \tfrac{1}{2} \times 506,$$

which checks.

Step 5. *U*-statistic of Group 1:
$$U_1 = 12 \times 10 + \tfrac{1}{2} \times 10 \times 11 - 81\tfrac{1}{2} \quad \ldots\ldots\ldots\ldots\ldots\ldots\text{Eq. 4.2}$$
$$= 120 + 55 - 81\tfrac{1}{2} = 93\tfrac{1}{2}$$

Step 6. *U*-statistic of Group 2:
$$U_2 = 10 \times 12 + \tfrac{1}{2} \times 12 \times 13 - 171\tfrac{1}{2} \quad \ldots\ldots\ldots\ldots\ldots\text{Eq. 4.3}$$
$$= 120 + \frac{156}{2} - 171\tfrac{1}{2}$$
$$= 26\tfrac{1}{2}$$

Step 7. Found value of *U* (the lesser of U_1 *and* U_2):
$$U_2 = 26\tfrac{1}{2}$$

Step 8. Arithmetic check by Eq. 4.4:
$$93\tfrac{1}{2} + 26\tfrac{1}{2} = 12 \times 10$$
$$120 = 120$$

which checks.

Step 9. Tabulated value of *U* for a 12, 10 comparison = 29.

Steps 10, 11. Since the found value of $U = 26\tfrac{1}{2}$ is *less than* the tabulated value of 29, the NH may be *rejected*, and the difference in medians declared significant at the $P = 5\%$ level.

For a full review of the analysis of survival times of animals in challenge experiments, see Liddell (1978).

4.5 Non-Parametric Testing of Paired Data

We have already discussed the analysis of paired data by *t*-test (Section 3.3.3). Now we come to the corresponding non-parametric procedures. The tests in question are Wilcoxon's Signed Rank test and Dixon and Mood's Sign test. They both differ from the *t*-test in that they require a minimum of 6 pairs, so we cannot apply them directly to the 4 pairs of blow-out/drainage data previously used for the *t*-test (Section 3.3.3). Moreover, if any of the pairs have measurements such that the difference within the pair is 0, additional pairs are required to give a minimum of 6 of what are called 'informative pairs', i.e. with non-zero differences.

The Signed Rank and Sign tests differ from each other in that the former is for use with (a) measurement data and (b) populations which have the same shape of distribution and differ only in the location of their medians. The Sign test is less restrictive, and is available for data of the type where one member of each pair is simply scored as 'greater than' or 'better than' the other, but with no attached measurements or assumptions about the shape of distribution. Different formulae and different statistical tables are used for the 2 tests, therefore they will be treated separately.

Table 4.2

(a) *Eight pairs of fictitious drainage/blow-out pipette deliveries for analysis by Wilcoxon's Signed Rank test.* (The same data are used also for Dixon and Mood's Sign test.)

Pair no.	1	2	3	4	5	6	7	8
Drainage (D)	7.6	2.6	4.6	11.7	10.8	12.7	13.0	9.7
Blow-out (B)	13.4	12.5	14.8	12.4	13.4	12.7	12.3	13.6
Difference (B − D)	5.8	9.9	10.2	0.7	2.6	0	− 0.7	3.9

(b) *Ranks of the (B − D) values, ignoring the zero value and treating + ve and − ve values as all + ve for purposes of ranking*

Pair no.	1	2	3	4	5	6	7	8
Rank of the the difference	5	6	7	1½	3	—	1½	4

4.5.1 Wilcoxon's Signed Rank test

Let us continue to work with blow-out/drainage pipetting data as our example, but this time let us suppose we have 8 pairs of results instead of the 4 pairs previously considered. As before, let us assume that we do not have information on whether it is legitimate to regard the data as samples from normal distributions and, through prudence, we are doing the analysis by a non-parametric method.

Table 4.2(a) sets out the fictitious blow-out/drainage data and also presents the difference between blow-out and drainage (B − D) for each pair. For pupposes of the Signed Rank test, zero values of (B − D) are regarded as worthless and we therefore have to discard the result of pair no. 6 (note that this discarding is not done in the *t*-test).

We then rank the difference values from the smallest upwards, but temporarily ignoring any minus sign and, at this stage in the calculation, basing the ranks purely on the magnitudes of the difference values. Ties are dealt with in the usual way, so that pairs 4 and 7 with values + 0.7 and − 0.7 are tied in ranks 1 and 2 and are each assigned a rank of 1½.

Next, we add up all the ranks associated with positive values of the difference and give this total the symbol T_+.

$$T_+ = 5 + 6 + 7 + 1½ + 3 + 4 = 26½.$$

Similarly, we add up all the ranks associated with negative values of (B − D) and give this total the symbol T_-.

$$T_- = 1½.$$

To check the accuracy of ranking and calculating T_+ and T_-, we make sure that

$$T_+ + T_- = ½N(N+1) \dots\dots\dots\dots\dots\dots\dots\dots\text{Eq. 4.5}$$

where N is the number of non-zero pairs:
$$26\frac{1}{2} + 1\frac{1}{2} = \frac{1}{2} \times 7 \times 8$$
$$28 = \frac{1}{2} \times 56$$
which checks.

For the null hypothesis (NH), we take as our found value of T whichever of T_+ and T_- *is smallest*, in this case $T_- = 1\frac{1}{2}$.

The NH states that 'the population medians of blow-out and drainage deliveries are equal'.

The NH is rejected if the *found T* is *less than or equal to the tabulated T*. Consulting Appendix A13 shows that for $N = 7$ non-zero pairs, the 5% point of T is 2. Therefore our found T of $1\frac{1}{2}$ represents a significant difference (just!) and leads us to conclude that there is a 'real' difference in the volumes delivered by drainage and by blow-out.

4.5.2 Dixon and Mood's Sign test

As indicated above, the Signed Rank test requires:

(a) that the data consist of measurements (and not just ranked observations), and

(b) that the two treatments, e.g. blow-out and delivery, represented in each pair, have similarly shaped underlying distributions, but may differ in locations of medians.

There are some types of observational data which cannot be expressed as actual measurements: examples are the intensities of local or general or histopathological reactions in animals or humans exposed to a noxious agent. In such instances it may only be possible to record that one member of the pair had a greater reaction than the other, but without being able to attach a measurement to the observation. To illustrate how the test is done, let us take the same blow-out/delivery observations as used for the Signed Rank test, but let us imagine that instead of having actual measurements of the weight of water delivered, we were only able to record which method gave the greater amount in each pair. As with the Signed Rank test, we ignore pairs where the members give indistinguishable values. Table 4.2 thus provides $n = 7$ *informative* pairs, in $r = 6$ of which, blow-out > drainage, and in $n - r = 1$ of which, drainage > blow-out.

The statistic for the Sign Test is R, defined as the smaller of r or $n - r$; in this case $n - r = 1$ is the smaller. We then look up the tabulated value of R (Table A14) for $n = 7$ and find the entry 0 for significance at the 5% level.

The NH states that 'the two populations represented in each pair are equivalent', and for the hypothesis to be rejected, the *found R* has to be less than or equal to the *tabulated R*. In this example, therefore, the result would be declared 'not significant'. This illustrates the general point that as the Tests of Significance shed their restrictions about the nature of the populations from which the data samples are drawn, so do they tend to lose in discriminating power. Thus, with the 8 pairs of observations under discussion, the unrestrictive

Sign test fails to find a significant difference between blow-out and drainage; the more fastidious Signed Rank test finds the difference significant at the 5% level. While if we assume that the data are normally distributed and analyse them by paired-sample t-test (Section 3.3.3) we find that the difference is significant almost at the 1% level. I am sure that any cynic reading these last few sentences may well feel that statistics can be used to prove whatever you want! But, of course, like all logical processes, the end-result depends on the initial assumptions.

4.6 Comparing Two Samples from Unknown and Not Necessarily Similar Distributions (K–S Test)

Let us get away from paired data and return to the comparison of two groups of measurements which we wish to analyse for significance of difference, but where we wish to avoid making any assumption about

(a) normality of the underlying distribution, or even

(b) similarity of shape of the underlying distributions from which the two groups came.

Condition (a) rules out the Student t-test and condition (b) rules out the Mann–Whitney U-test.

One might think that the experimental biologist would make extensive use of such an unrestrictive test of significance, but this does not appear to be so. Perhaps it has been avoided because it goes under the ponderous name of the Kolmogorov–Smirnov test. For brevity we shall refer to it as the K–S test. Although not requiring assumptions (a) and (b) above, the K–S test *does* have the restriction of needing equal numbers of observations in the 2 groups, and a group size of at least 4. A major feature of the test is its sensitivity to differences in both the shape of the distribution in the underlying populations, and to the location of the medians. Thus, a significant difference between groups in the K–S test could arise from either or both effects, but the K–S test does not tell you which.

4.6.1 Conduct of the K–S test

The K–S test is simple and does not involve any formulae or calculations (so why is it not more popular?). As data for a worked example, let us take the 16 blow-out/drainage pipette measurements used in the previous section, but simply ignore the pairing, i.e. let us imagine we have 8 randomly-obtained blow-out deliveries and 8 randomly-obtained drainage deliveries. For convenience, let us take numerical values the same as those already used in the discussion on paired data. This, incidentally, is not being done through laziness or lack of imagination on the part of the author, nor because the author thinks that experimental biologists spend their time weighing drainage and blow-out deliveries from pipettes. No! The idea is to show how the *same* set of figures can be analysed statistically in a variety of ways, often leading to different conclusions, depending on the exact circumstances under which the data were gathered and the assumptions made about their nature. It is to drive home one of the central messages of this book, namely that the statistical analysis of an experiment, or survey, must be tightly integrated with the layout and conduct of the actual practical work and vice versa. *Experiment-plus-analysis should make up one integrated package.*

The first step in the K–S test is to arrange the data in each group in ascending numerical order. This gives us 'working mg' values as follows:

Drainage: 2.6, 4.6, 7.6, 9.7, 10.8, 11.7, 12.7, 13.0
Blow-out: 12.3, 12.4, 12.5, 12.7, 13.4, 13.4, 13.6, 14.8

We then take the measurement scale of 'working mg' and divide it up into equal intervals so as to accommodate all the readings from 2.6 to 14.8. We should choose an interval width such that not more than 2 or 3 readings fall into a given interval, or the test loses sensitivity. For convenience initially, let us take a 1-mg-wide interval, so that a scale with upper boundaries from 3 to 15 will accommodate all the data. Later we can see if a 0.5-mg interval would be better, and if a 2-mg-wide interval would have been unsatisfactory.

As shown in Table 4.3, we enter in columns 2 and 3 the drainage and blow-out measurements in their appropriate place opposite the interval scale in column 1. We then, in columns 4 and 5, progressively add up the entries so as to get cumulative frequencies from 0 to 8. The right-hand column, headed Difference, is the key one. The vertical bars around Difference simply mean that we subtract

Table 4.3   Kolmogorov–Smirnov (K–S) test applied to fictitious blow-out/drainage pipetting data

Upper boundary of interval (mg) (1)	Measurement data		Cumulative frequencies		
	Drainage (2)	Blow-out (3)	Drainage (4)	Blow-out (5)	Difference (6)
3	2.6	—	1	0	1
4	—	—	1	0	1
5	4.6	—	2	0	2
6	—	—	2	0	2
7	—	—	2	0	2
8	7.6	—	3	0	3
9	—	—	3	0	3
10	9.7	—	4	0	4
11	10.8	—	5	0	5
12	11.7	—	6	0	6
13	12.7 13.0	12.3, 12.4 12.5, 12.7	8	4	4
14	—	13.4, 13.4 13.6	8	7	1
15	—	14.8	8	8	0

the drainage and blow-out frequencies on each line in either direction so as to get a positive difference. We then identify the *maximum* difference in frequency, in this case 6.

This maximum frequency difference of 6 is our *found value* of the K–S statistic. We then look up the K–S Table (Appendix A15) and find that for a group size of 8, a K–S value of 5 is significant at the $P = 5\%$ level and a K–S value of 6 is significant at the 1% level. We can therefore conclude that our drainage and blow-out deliveries are significantly different at the 1% level, i.e. are highly significantly different either in the value of the medians or in the degree of scatter—the test does not tell us which. Looking at the distributions of the data as displayed in Table 4.3, it would appear that there are differences both in the shapes of the two underlying populations and in the location of their medians. The drainage values are much more highly scattered from 2.6 to 13 mg, as well as having a lower median value (between 9.7 and 10.8 mg); whereas the blow-out values are tightly bunched between 12.3 and 14.8 mg, with a median between 12.7 and 13.4 mg.

If we had gone ahead blindly and done a t-test on these data, we would have got the following results:

Drainage: $\bar{y} = 9.087$, $s = 3.834$, $s^2 = 14.704$
Blow-out: $\bar{y} = 13.137$, $s = 0.842$, $s^2 = 0.7084$

Substituting the $\bar{y}$ and s values into Eq. 3.1 yields $t = 2.918$ with 14 d.f. This corresponds to P between 1 and 2%. Note, however, that the t-test is not justifiable because of the marked difference in variances of the two groups. The variance ratio

$$F = \frac{S_D^2}{S_B^2} = \frac{14.704}{0.7084} = 20.75.$$

Consulting the F-table (A3) shows that this corresponds to a highly significant F-ratio (tabulated F for d.f. 7,7 and $P = 1\%$ is 6.99).

In a 'real life' investigation, where one is generating results in the same general format as those in Table 4.3, one would never take a single day's work as definitive. Instead, one would repeat each experiment several times and base one's decision on which statistical test to use from the accumulated experience. It might transpire, for example, that the data of Table 4.3 were unrepeatable and that in all other runs of the experiment the data could confidently be analysed by the t-test. On the other hand, it might turn out that drainage persistently is more variable than blow-out, and that the K–S test would be safer. In conclusion, the reader may note that the Behrens–Fisher test, which is not dealt with in this book but can be found in Campbell (1974, p. 158), is available for comparing the means of samples from two normally-distributed populations with different variances.

Table 4.4 Tests for analysis of grouped data from unknown distributions

Test	Purpose	Requirements/features
Rankit	To see if the assumption of normality is justified	$N \geqslant 5$
Mann–Whitney U-Test	To compare the medians of 2 groups	1. $N \geqslant 4$ 2. Distributions assumed to be of same shape
Wilcoxon's Signed Rank test	To compare medians in paired data	1. No. of pairs $\geqslant 6$ 2. Distributions assumed to be of same shape 3. Data consist of measurements
Dixon and Mood's Sign test	To compare merely-ranked paired data	1 and 2 as above, but pairs can be scored as 'greater' or 'better'
Kolmogorov–Smirnov test	To compare 2 groups for difference in either medians, or shape of underlying distribution	1. $N \geqslant 4$ 2. Same no. of observations in the 2 groups 3. Test does no discriminate between difference in medians or shapes
Friedman S-test	To compare the medians of up to 5 groups	1. Data must be in randomized block design 2. Distributions assumed to be of same shape
Kruskal–Wallis H-test	To compare the medians of 3 groups	1. Groups may be of unequal size 2. Distributions assumed to be of same shape 3. Can be extended to >3 groups

4.6.2 Choice of interval in the K–S test

In the evaluation of the above example, a 1-mg-wide class interval was chosen rather arbitrarily, and as it happened, revealed a highly significant difference between the two groups. We did, however, violate the normal requirement that the interval width should be chosen so as not to have more than 2 or 3 observations of either group in an interval. Taking a 0.5-mg interval does not, however, materially alter the outcome of the analysis except in reducing the peak occupancy from 4 results to 3. The maximum difference in cumulative frequency remains at 6. Likewise, increasing the interval width to 2 mg has only a slightly deleterious effect on the analysis in reducing the maximum cumulative frequency difference to 5. It therefore seems that our initial choice of the 1-mg interval-width was satisfactory.

4.7 Non-Parametric Testing of Three or More Samples and General Summary

The Mann-Whitney U-test for comparing the medians of two groups of data cannot directly be extended to comparing the medians of 3 or more groups. There are, however, two similar non-parametric tests available for this latter purpose, namely the Friedman test and the Kruskal–Wallis test, the former of which is described in Chapter 8.

A general summary of the non-parametric tests for analysis of grouped data is given in Table 4.4. Useful articles on the application of non-parametric tests respectively to microbiological and biochemical data have been provided by Jones (1973) and Nimmo and Atkins (1979).

5
How to Deal
with Proportion Data

Life is short, Art long, Opportunity fleeting, Experience treacherous, Judgement difficult.

Hippocrates, Aphorisms.

5.1 Developing a Sense of Proportion

What may conveniently be called *proportion data* can arise from a wide variety of procedures such as therapeutic trials of drugs and vaccines, animal toxicity, infection and immunization studies, water microbiology, the sterility testing of injectable medicines, assay of viruses in roller tubes or microtitre plates, and genetical experiments in animal and plant breeding. The feature common to all of these is that the raw results consist of whole-number proportions, such as 3/8 or 41/375. As pointed out earlier, you cannot have 7.3 mice alive in a cage after giving 12 animals a challenge dose of a virulent organism. Nor in a breeding experiment can you have a litter with 3.63 females and 5.37 males. The essence of all such data is that each *unit* (mouse, test tube, etc.) under observation is scored as + or − , or as female or male, dead or alive, turbid or clear, responding or not responding, etc. These are what the statisticians call *classificatory data*—data in which each experimental object is unambiguously assigned to one of two categories or classes. Usually it is 2 classes, but it does not have to be. For example, an ordinary die has 6 possible categories of result.

Having got such data, what are the things we usually want to do with them? Among the possibilities are:

(1) Analysis of *differences* in drug trials, etc. Is treatment A significantly better than treatment B?

(2) In sterility testing of injectable drugs, calculating the probability of a 'clear' test on a sample of $N = 20$ ampoules if the background incidence of contamination in the whole production batch is $X\%$.

(3) Estimation of LD_{50}, ED_{50} values. What dose of drug, toxin, challenge organisms, etc. can be interpolated as affecting 50% of the animals, plants, etc.?

(4) In water analysis by the Most Probable Number (MPN) method, what is the viable count of organisms in the test sample?

(5) In animal and plant genetical experiments, do the observed ratios of phenotypic characters in the progeny correspond to those expected from Mendelian principles?

From a procedural point of view, the statistical approach to proportion data depends on how many observations we have. Is it a *large* experiment or a *small* one? Are we dealing with groups of 5 or 10 mice or broth tubes, or with scores, hundreds or thousands of individuals in a drug trial? The exact difference between large and small will emerge in Section 5.2.4. However, as an initial rough guide, we can take any group of 20 or less as being 'small', and anything of several score or more as being 'large'. Let us now deal with the 5 types of question posed above.

As an aside to the reader who is looking for methods for dealing with antibody titration data, where the titre may be expressed as a proportion such as 1/128, note that such data are best normalized by expressing as the logarithm of the reciprocal, e.g. 1/128 is converted to 128 then expressed either as the $\log_{10}$ value 2.107 or the $\log_2$ value 7.0.

5.2 Is that Difference Significant?

5.2.1 Comparing two small groups

Let us take as our example an animal challenge experiment with 2 groups of 10 mice in which the proportions dying after receiving different treatments A and B were 4/10 and 8/10, respectively. Question: Do these results indicate that Group B has a significantly higher mortality than Group A? Answer: No. How do we arrive at this assessment? Well, we can either make calculations based on the binomial distribution formula, which are rather tedious, but which we shall do nevertheless — but not straightaway — or we can take the view that 'there is no point in having a dog and then doing your own barking', because Finney *et al.* (1963) have produced a series of tables in a book called *Tables for Testing Significance in a 2 × 2 Contingency Table* where all the work is already done. This book provides all the permutations and combinations of group size and results for groups with up to 40 in each. So that, for example, you can compare 3/18 with 9/39 by entering the table at the appropriate page. Obviously this whole book cannot be reproduced here. Instead, a few commonly-used group sizes have been selected. Furthermore, the tabular information of Finney *et al.* (1963) has been recast as diagrams (Appendix A11) for visual convenience. Let us now go back to the results 4/10 and 8/10 and consult the 10 × 10 diagram of Appendix A11 which, for convenience, is reproduced as Fig. 5.1.

Step 1. With either fraction, e.g. the 4/10, enter the numerator 4 on the top horizontal line of the diagram, as shown by the arrow.

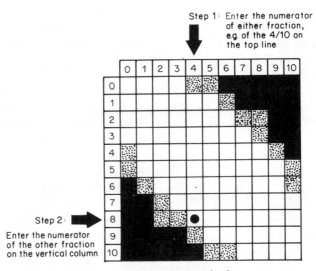

Fig. 5.1 Diagram for assessing the significance of differences in 2 groups of 10 observational units in each of which there is an all-or-nothing response. Example: is 8/10 significantly higher than 4/10? After the 3-step procedure, the interpretation is dependent upon the location of the point of intersection:

clear: result not significant ($P > 5\%$);
dotted areas: significant at the $P = 5\%$ level;
dark areas: significant at the $P = 1\%$ level.

Therefore 8/10 is not significantly higher than 4/10.
Note: this is a one-tail test.

Step 2. With the other fraction, 8/10, enter its numerator 8 in the left-hand vertical column.

Step 3. Determine the point of intersection of the two entries.

Step 4: Interpretation. If, as here, the point of intersection is in the central white area of the diagram, the result is interpreted as *not significant* (at the 5% probability level). If the intersection point is in the dotted area, the result is *significant* at the $P = 5\%$ level, and if it lies in the dark area it is significant at the $P = 1\%$ level.

Thus, Group B with 8/10 does not have a significantly higher mortality than Group A with 4/10. If the results had been 3/10 and 8/10, then the difference would have been significant at the 5% level, as would 2/10 and 8/10; while the difference between 1/10 and 8/10 would be judged significant at the 1% level.

As with all tests of significance, the above procedure involved setting up a null hypothesis, although this has not so far been stated explicitly. This hypothesis is that 'there is no evidence of a higher death rate in Group B, than in Group A, other than what can readily be accounted for by random sampling fluctuations'. As with other tests of significance, we use the convention of rejecting the null hypothesis if the probability of the result occurring by chance is 5% or less. Here the probability is greater than 5% and therefore the difference is judged as not significant.

Two additional points about this table:

(1) It does not matter which fraction is entered on the top horizontal line or left-hand vertical column, because of the symmetry of the diagram.

(2) The test of significance used here in a *one-tail test*, i.e. we ask not simply whether the mortality rates in the 2 groups are *different*, but more specifically whether Group B has a significantly *higher* mortality rate than Group A. This is useful for experiments that involve comparing a Treatment Group with a Control Group since usually the focus of interest is whether the treatment result is specifically higher or lower than the control and not just whether it is different.

5.2.2 What to do if 'Finney-less'

Those accustomed to biological work know that because of the operation of 'Murphy's Law' (see Glossary), experiments that start with equal-sized groups do not always finish up this way; so that when the results come to be analysed there may be awkward combinations of proportions such as 3/9 to be compared with 8/11. In the absence of the tables of Finney *et al.*, the following procedure may be followed. For purposes of instruction we shall not use 3/9 and 8/11 as the example for working. Instead, it will be more illuminating to take the results 4/10 and 8/10 and work through them by the alternative method to show that the conclusions agree with those reached by the use of Fig. 5.1.

Step 1. Express the data as a *2 × 2 contingency table*, that is, a table with 2 horizontal rows and 2 vertical columns which make 4 slots into which the observed numbers may be inserted. The table also has row-totals and column-totals. So the results 4/10 and 8/10 may be presented as:

Group	Responding	Not responding	Total
A	4	6	10
B	8	2	10
Total	12	8	20

The format of such a 2×2 contingency table may be generalized by the use of symbols:

Group	Responding	Not responding	Total
A	a	b	$(a+b)$
B	c	d	$(c+d)$
Total	$(a+c)$	$(b+d)$	N

a, b, c and d can have any whole-number values.

Step 2. Set up a null hypothesis which states that there is no difference between the results from Groups A and B, and therefore any higher mortality in B is due purely to random-sampling fluctuations. Effectively this mean that we can lump together 2 groups of 10 mice and think of them as a single group of 20 animals, of which 12 responded and 8 did not. This is mathematically equivalent to having a sack with 12 black balls and 8 white ones. We reach blindly into the sack and take 10 balls at random, leaving 10 behind. What is the chance that the 2 sets of 10 will contain 8 black in the one and 4 black in the other? Does this combination have such a high chance of occurrence that it would be a common event in random-sampling operations? This is what we now have to investigate.

Step 3. Express the 4 numbers in the central slots of the 2×2 contingency table in a summarized form as the array:

$$\frac{4 \mid 6}{8 \mid 2} \text{ which may be generalized as } \frac{a \mid b}{c \mid d}$$

Step 4. Construct *all the other possible arrays* of the numbers in the central slots *which are permissible without changing either the row or the column totals.* These other permissible arrays are:

$$\frac{3 \mid 7}{9 \mid 1} \quad \frac{2 \mid 8}{10 \mid 0} \quad \frac{5 \mid 5}{7 \mid 3} \quad \frac{6 \mid 4}{6 \mid 4} \quad \frac{7 \mid 3}{5 \mid 5} \quad \frac{8 \mid 2}{4 \mid 6} \quad \frac{9 \mid 1}{3 \mid 7} \quad \frac{10 \mid 0}{2 \mid 8}$$

So we have a total of 9 permissible arrays, including the one given by our experimental samples. No other arrays are possible without alteration of marginal totals. Check that each of the above arrays has row-totals of 10 and 10, and column totals of 12 and 8.

Step 5. Calculate the probability (*P*) of *each* of the above arrays by applying the formula:

$$P(\%) = \frac{100(a+b)!(c+d)!(a+c)!(b+d)!}{N!a!b!c!d!} \quad \ldots\ldots\ldots\text{Eq. 5.1}$$

So that for our experimental sample, we have:

$$P(\%) = \frac{100 \times 10! \times 10! \times 12! \times 8!}{20! \times 4! \times 6! \times 8! \times 2!}$$

(where $10! = 10 \times 9 \times 8 \times 7 \times 6 \times 5 \times 4 \times 3 \times 2 \times 1 = 3,628,800$)

$$P(\%) = 7.502$$

That is, a probability of 7.5%. (Note that the above calculation can be done quite easily even with a calculator that lacks a factorial button, by careful cancellation of the factorial quantities on the top and bottom lines.) Now work out the *P*-values for all of the other arrays, but note two labour-saving devices:

(1) The top line of the equation divided by *N*! is a constant for all of the arrays. Therefore it should be worked out separately and stored in the memory of the calculator.

(2) Four of the arrays are arithmetically identical with four others, differing merely in which members are allocated to which letters in the formula. Thus, having worked out $\dfrac{4 \mid 6}{8 \mid 2}$, there is no need to do a separate calculation of $\dfrac{8 \mid 2}{4 \mid 6}$ except to check arithmetic.

Step 6. We can now tabulate the series of arrays and their associated *P*-values, and also set down the proportions which are being compared, but in the usual way of recording, i.e. 4/10, etc. This gives the results in Table 5.1. Inspection of the bottom line of the table reveals a regular probability distribution (Fig. 5.2) with a maximum of 35% in the middle, falling to 0.036% at the two ends. Note that the sum of the probabilities is 100% — which is a useful check on the arithmetic.

Step 7. This is the only part of the procedure which is likely to cause difficulty. Referring to set 3 in Table 5.1, we see that this set has a probability of 7.5%, i.e. we can expect it to arise on 7.5% of occasions during a long run of random sampling. However, in order to make the one-tail test of significance, the probability that we take for judging the null hypothesis is not just the 7.5% but also the *P*-values of 0.036% and 0.96% *which make up the rest of that tail* of the probability distribution. Therefore when we fail to reject the null hypothesis, it is on the basis of a probability of $7.50 + 0.96 + 0.036 = 8.496\%$.

Table 5.1 A list of comparisons, arrays and probabilities for the 'Finney-less' method of determining the significance of the difference between 4/10 and 8/10

Set no.	1	2	3	4	5	6	7	8	9
Comparison	10/10 versus 2/10	9/10 versus 3/10	8/10 versus 4/10	7/10 versus 5/10	6/10 versus 6/10	5/10 versus 7/10	4/10 versus 8/10	3/10 versus 9/10	2/10 versus 10/10
Array	$\begin{array}{c c} 10 & 0 \\ \hline 2 & 8 \end{array}$	$\begin{array}{c c} 9 & 1 \\ \hline 3 & 7 \end{array}$	$\begin{array}{c c} 8 & 2 \\ \hline 4 & 6 \end{array}$	$\begin{array}{c c} 7 & 3 \\ \hline 5 & 5 \end{array}$	$\begin{array}{c c} 6 & 4 \\ \hline 6 & 4 \end{array}$	$\begin{array}{c c} 5 & 5 \\ \hline 7 & 3 \end{array}$	$\begin{array}{c c} 4 & 6 \\ \hline 8 & 2 \end{array}$	$\begin{array}{c c} 3 & 7 \\ \hline 9 & 1 \end{array}$	$\begin{array}{c c} 2 & 8 \\ \hline 10 & 0 \end{array}$
$P(\%)$[a]	0.036	0.96	7.50	24.01	35.01	24.01	7.50	0.96	0.036

[a]Note that these add up to 100.0 to within the accuracy of rounding-off errors.

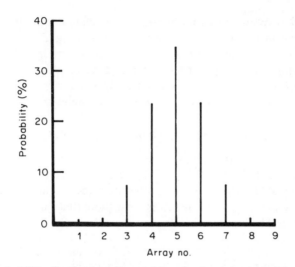

Fig. 5.2 Probability-distribution diagram for the 9 arrays in Table 5.1. Since the distribution is inherently discontinuous, the probabilities are represented as thin vertical lines rather than as the connected blocks which would be appropriate for the histogram of a normal distribution

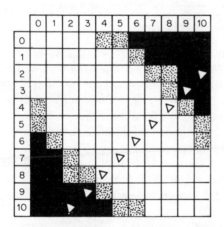

Fig. 5.3 The 9 arrays of contingency data of Table 5.1 shown superimposed on Fig. 5.1

This is the probability which leads to the conclusion that 8/10 is not significantly higher than 4/10.

A similar pattern exists at the other end of the distribution where sets 7, 8 and 9 taken together also have a summed probability of 8.496%.

Thus, in the *one-tail* test of significance as done above, we are unable to reject the null hypothesis because the results have a $P = 8.496\%$. In a two-tail test, the combined probabilities in the two tails would be 17%, which would be even further away from being a significant difference.

The set of 9 arrays whose probabilities are calculated in Table 5.1 have been inserted as the row of triangle-containing squares in Fig. 5.3. This gives some idea of the amount of calculation involved in producing Fig. 5.1, i.e. a 10×10 table has 100 pigeonholes in it, each of which requires substitution of a set of numbers into Eq. 5.1. As already emphasized, it is worthwhile to get a copy of Finney's tables if you generate such data routinely!

5.2.3 The prevention of rabies by vitamin C: use of the χ^2 test

As an introduction to the χ^2-test, the working example is the experiment described by S. Banic (Nature **258**, 153–4, 1975) in which injections of vitamin C were investigated for their possible protective effect in guinea pigs which had been given an LD_{50}–LD_{90} dose of rabies virus (note that this is not a very large challenge dose). The results were:

Treatment	No. of deaths	No. of survivors	Total
Vitamin C	17	31	48
Control	35	15	50
Total	52	46	98

Reproduced by permission from Nature, *258, 153. Copyright © 1975 Macmillan Journals Limited.*

The first thing to note is that 'large-group' proportion data such as the above are easier to analyse than the 'small-group' data we have just been considering. In Section 5.2.4 we shall see exactly where the boundary between 'large group' and 'small group' lies. Meanwhile, let us look at two approaches to the analysis of the vitamin C results. They give almost the same answer and the reason for using them both is because some useful statistical insights will be gained thereby. First the simplest approach.

Method 1, Step 1. Regard the data as fitting into a 2×2 contingency table with data slots a, b, c and d, and with marginal totals $(a+b)$, $(c+d)$, etc. as in Step 1 of Section 5.2.2. We therefore have $a = 17$, $b = 31$, $c = 35$, $d = 15$, and $N = 98$.

Step 2. Substitute the above values into the formula

$$\chi^2 = \frac{N\{|ad - bc| - \frac{1}{2}N\}^2}{(a+b) \times (c+d) \times (a+c) \times (b+d)} \quad \ldots\ldots\ldots\ldots \text{Eq. 5.2}$$

The straight vertical brackets on the top line indicate that the subtraction of the products ad and bc is to be done so as to yield a positive answer.

Substituting, therefore, we get

$$\chi^2 = \frac{98\{|17 \times 15 - 31 \times 35| - \frac{1}{2} \cdot 98\}^2}{48 \times 50 \times 52 \times 46}$$

$$= 10.41.$$

This is the *found value of* χ^2 and is thus analogous to the *found value of t* or the *found value* of the various other statistics which were dealt with in Chapters 3 and 4.

Step 3. Compare the *found* value of χ^2 with the *tabulated* value of χ^2. To do this we need to be able to enter the χ^2 table (Appendix A4) with the appropriate number of degrees of freedom. For *any* 2×2 contingency table, the d.f. = 1. This is irrespective of the size of the numbers in the 2×2 table. The χ^2-table for 1 d.f. gives the following entries:

$P(\%)$	10	5	1	0.1
χ^2	2.706	3.841	6.635	10.827

Our found value of $\chi^2 = 10.41$ therefore has a *P*-value almost as low as 0.1%, i.e. the experimental results have only about a 0.1% probability of occurring as a random-sampling fluctuation, assuming the null hypothesis to be correct. Therefore, by convention, we would judge the effect of vitamin C to be *statistically highly significant.*

Let us now use the χ^2-test on the same data, but adopting a slightly different approach. This is done mainly to introduce the concepts of *expected* numbers and *observed* numbers.

Method 2, Step 1. Set out the results as a 2×2 contingency table as before. The four numbers in the central pigeonholes of this table are the *observed numbers* (O).

Step 2. For each of the four observed numbers (O), calculate a corresponding *expected number* (E). These expected numbers are calculated *after* making the assumption that vitamin C has no effect, i.e. we are setting up a null hypothesis and then making calculations on the basis of it being correct.

Thus, of the 98 animals in the trial, a total of 52 died. If, therefore, we were to take a random sample of 48 animals (the number in the Vitamin C Group) we would *expect* the number of deaths, by simple proportion, to be

$$48 \times 52/98 = 25.4694.$$

Likewise in the Control Group where there were 50 animals, we would *expect* $50 \times 52/98 = 26.5306$ deaths.

We then make similar calculations for the expected numbers of *survivors*. The overall survival rate was 46/98. Therefore in a random sample of 48 animals (the number in the Vitamin C Group) we would expect $48 \times 46/98 = 22.5306$ survivors. And in the Control Group of 50 animals we would expect $50 \times 46/98$ survivors.

Note that in these calculations we give the expected numbers with several decimal places if need be, because there is no theoretical requirement for these expectations, which reflect long-run probabilities, to work out as whole numbers.

Step 3. Draw up a table to present the observed and expected numbers for each pigeonhole in the 2×2 contingency table. Then calculate $(O - E)$ and $(O - E)^2/E$, as shown in Table 5.2.

Table 5.2 Observed and expected numbers for the rabies/vitamin C experiment, and calculation of χ^2

Treatment category and response	Observed number (O)	Expected number (E)	$\|O-E\|$	$\dfrac{(O-E)^2}{E}$
Vitamin C—died	17	25.4694	8.4694	2.8163
Vitamin C—survived	31	22.5306	8.4694	3.1837
Control—died	35	26.5306	8.4694	2.7037
Control—survived	15	23.4694	8.4694	3.0564
Total	98	98.0000	—	11.7601

Step 4. Add up the values of $(O-E)^2/E$ in the last column of Table 5.2 to give $\Sigma(O-E)^2/E$. We then use the simple formula:

$$\chi^2 = \Sigma(O-E)^2/E \dotfill \text{Eq. 5.3}$$

$$= 11.76.$$

This is our *found* value of χ^2 which we then compare with the tabulated values of χ^2 for d.f. $= 1$, as done in the first method. As before, we find that the result is highly significant, although the found value of χ^2 is slightly different from that yielded by Method 1. So which of the two *found* values of χ^2 is correct? The answer is that they are both 'correct', but the one given by Eq. 5.2 is a 'better' value. The reason for this is that Eq. 5.2 contains on the top line the term $1/2\,N$ which is known as 'Yates' correction'. So why did we bother to go through Method 2 when Method 1 gives a more accurate *found* value of χ^2? The answer is to introduce the concepts of *observed* and *expected* numbers which are needed in the next section.

5.2.4 The meaning of 'small group' in the present context

It was pointed out at the beginning of this chapter that the choice of method for the analysis of the difference between 2 groups of proportion data depended on whether the group size was 'large' or 'small'. For introductory purposes, it was suggested that groups of 20 or less could be considered 'small'. Now, having introduced the concept of *expected numbers*, we are able to give more exact instructions. The rule is that the χ^2-test with formulae as given in Eqs 5.2 and 5.3 should only be used provided that none of the four *expected* numbers of the 2×2 contingency table is less than about 5. So that with groups of 20, for example, it would be just permissable to use χ^2 if the two proportions were 1/20 and 9/20, because the smallest expected number would be $20(1+9)/40 = 5$. On the other hand, it would be bending the rule to compare 0/20 with 6/20, because the smallest expected number would be $20(0+6)/40 = 3$.

With groups much smaller than 20 it becomes progressively harder to avoid having at least one expected number less than 5. With groups of 10 it is impossible, except for the single comparison of 0/10 and 10/10.

So although at the beginning of this chapter the difference between a 'large group' and a 'small group' was given in terms of total group size, the more exact criterion is that *no expected number* in the 2×2 contingency table should be less than 5.

5.3 Are Those Ampoules Sterile?

In the pharmaceutical industry, sterility tests are performed to ensure that products labelled 'sterile' are free from living microbes. Once a large batch of an injectable substance has been dispensed into its final ampoules or vials, the law requires that 2%, or 20, of the containers, whichever number is smaller, should be tested for sterility by inoculation into suitable culture media (*British Pharmacopoeia*, 1980). What concerns us here is to calculate the probability that a 20-ampoule sterility test might be passed as satisfactory if $X\%$ of the whole batch of ampoules were in fact contaminated. Or, looking at it the other way, if a sample of 20 ampoules shows *no* evidence of contamination, what is the probability that $X\%$ of the batch from which the sample was drawn *might* be contaminated. This problem also permits us to introduce the binomial distribution which is one of the fundamental distributions in statistics, ranking with the normal distribution in its theoretical importance.

The solution to the problem posed is as follows

Step 1: Introduction of symbols. Let X, expressed as a decimal of 1.0, be the proportion of *contaminated* ampoules in the whole production batch, so that if, for example, 1% of the ampoules were contaminated, X would be 0.01. $(1 - X)$ is therefore the proportion of *sterile* ampoules in the whole production batch $= 0.99$. Normally, of course, one would not know what the value of X is, so we are dealing with a hypothetical case. Let $N =$ the number of ampoules taken for the sterility test. Usually $N = 20$. Let $a =$ the number of contaminated ampoules found in the sample of size $N = 20$ during the actual sterility test; a can have any whole-number value between 0 and 20.

Step 2: Introduction of formula. The probability (P_a) of finding a contaminated ampoules in a sample of size N taken randomly from a production batch in which the decimal fraction X of the ampoules are contaminated is given by the binomial formula:

$$P_a = \frac{N!}{a!\,(N-a)!}\,(X)^a \cdot (1-X)^{N-a} \quad \ldots\ldots\ldots\ldots \text{Eq. 5.4}$$

Step 3: Insertion of numbers. Let us suppose that 1% of the ampoules are contaminated. What is the chance of failing to detect this (i.e. put $a = 0$) in a sample of size $N = 20$ ampoules?

$$P_{a=0} = \frac{20!}{0!\,(20-0)!} \cdot (0.01)^0 \cdot (1-0.01)^{20}$$

A lot of this cancels: O! is equal to 1 (this follows from putting $N = 1$ into $N!/(N-1)! = N$); and $(0.01)^0$ is equal to 1.0. So we are left with:

$$P_{a=0} = (0.99)^{20}$$
$$= 0.82$$

This means that with a 1% contamination rate in a production batch, there is a probability of 82% that the sterility test would be recorded as satisfactory and only a chance of 18% that one or more contaminated ampoules would be detected in the 20-ampoule sterility test.

A contamination rate of 1% is quite unreasonably high for a modern ampoule-filling machine when used correctly and a rate of 0.1%, or 1 contaminated ampoule per 1,000 filled, might be more realistic. The chance of failing to detect contamination at this level in a 20-ampoule test is:

$$P_{a=0} = (0.999)^{20}$$
$$= 0.98$$

i.e. the chance of *actually detecting* 0.1% contamination by a 20-ampoule test is only 2%, which means that the manufacturer should not place much reliance on the 20-ampoule sterility to detect low incidences of contamination. It might, however, detect a gross inadequacy in the manufacturing procedure, such as failure to sterilize a batch of empty ampoules before filling them. For example, if 50% of the ampoules were contaminated, the chance of *not* detecting this is:

$$P_{a=0} = (0.50)^{20} = 0.00000095$$

which means that the probability of *detecting* it is $1 - 0.00000095 = 0.99999904$, or 99.999904%.

So the 20-ampoule sterility test is to be regarded as a spot check to pick up a serious failure in the manufacturing process. Note also that this ampoule example has its counterpart in the food industry when testing canned foods.

5.4 Estimation of ED_{50}, LD_{50} and Their Confidence Limits

Among the common laboratory procedures which generate proportion data are animal experiments set up to titrate, for example, the toxicity of bacterial toxins,

Reproduced by permission of *New Scientist*, the weekly review of science and technology

the virulence of live microorganisms, the protective efficacy of drugs and vaccines and similar operations in which each experimental unit (animal) is scored as 'responder' or 'non-responder', or as 'dead' or 'alive', etc. Having performed such titrations, one usually wants some simple expression for the potency of the administered material, and the ED_{50}, or dose that effects the change in 50% of the animals, is commonly the most useful index. If the response of the animal is death, then the initials ED_{50} are made more specific by redesignation as LD_{50} for the 50% lethal dose. Occasionally it may be appropriate to determine the LD_0 or LD_{100}, i.e. the highest dose that kills none, or the lowest dose that kills all, of the animals; but, statistically these are much less satisfactory end-points because of the sigmoid nature of the dose–response curve, e.g. Fig. 5.4 below, and the impossibility of exact interpolation in the asymptotic regions.

There are several methods for determining the ED_{50} from a set of data (e.g. see Finney, 1971, p. 437). However, we shall consider only two, namely the Spearman–Kärber and the probit methods. As we shall see, the latter is the most generally useful, versatile and efficient.

A few decades ago a popular method for finding ED_{50}s was that of Reed and Muench, and frequent reference will be found to it in the older literature. However, it is no longer considered statistically respectable and will not therefore be given here. A further procedure is the *moving average method* of Thompson, which is well described by Meynell and Meynell (1970, p. 207). This method is not given here (a) because of the extensive tables needed with it and (b) because it does not accommodate results with unequal numbers in the denominators of the proportions.

5.4.1 Spearman–Kärber method

This method has the merits of simplicity and statistical respectability but it has the significant drawbacks of requiring:

(1) even spacing between the dilutions (not much of a drawback);

(2) the same number of units (e.g. animals) set up at each dose level;

(3) a sufficiently broad range of dilutions to yield groups showing 0% and 100% of responders (this cannot always be guaranteed).

Let us illustrate its use with the following set of data from the virulence titration of a bacterial suspension inoculated into mice:

Dilution of challenge suspension	10^{-2}	10^{-3}	10^{-4}	10^{-5}	10^{-6}	10^{-7}
No. dead/ no. challenged	15/15	14/15	13/15	9/15	1/15	0/15

The symbols needed are:

m = the $\log_{10}$ dilution corresponding to the largest proportion. This latter is the 15/15 result given by the 10^{-2} dilution, and the $\log_{10}$ value of this dilution is -2, which is therefore m;

Δ = the constant interval between the $\log_{10}$ dilutions = 1;

P = the mortality fraction at each dilution expressed as a decimal. This gives p-values from left to right as: 1.0, 0.933, 0.867, 0.6, 0.067, 0.

We then apply the formula:

$$\log_{10} ED_{50} = m - \Delta(\Sigma p - 0.5) \quad\dots\dots\dots\dots\dots\dots\dots\text{Eq. 5.5}$$

where

$$\Sigma p = 1.0 + 0.933 + 0.867 + 0.6 + 0.067 = 3.467$$

Substitution into Eq. 5.5 gives:

$$\begin{aligned}
\log_{10} ED_{50} &= -2 - (1)(3.467 - 0.5) \\
&= -2 - 2.967 \\
&= -4.967
\end{aligned}$$

The ED_{50} is thus $10^{-4.967}$ or a dilution of 1/92,680 (i.e. the reciprocal of the antilog$_{10}$ of -4.967. Let us suppose that the undiluted bacterial suspension contained 6×10^7 colony-forming units (CFU) in the challenge dose. Thus, the LD_{50} in terms of CFU is given by $6 \times 10^7 \div 92,680 = 650$ (to 2 significant figures).

We can then go on to calculate the standard error of this estimate of the ED_{50}, as defined by:

$$S_{\log_{10} ED_{50}} = \Delta \sqrt{\frac{\Sigma p(1-p)}{(n-1)}} \quad\dots\dots\dots\dots\dots\dots\dots\text{Eq. 5.6}$$

where n is the number of units (animals) at each dilution. Here $n = 15$.

$p(1-p)$ is calculated for each dilution as follows:

Dilution	10^{-2}	10^{-3}	10^{-4}	10^{-5}	10^{-6}	10^{-7}
Dead/challenged	15/15	14/15	13/15	9/15	1/15	0/15
Proportion (p)	1.0	0.933	0.867	0.6	0.067	0
$(1-p)$	0	0.067	0.133	0.4	0.933	0
$p(1-p)$	0	0.0625	0.1153	0.24	0.0625	0

Thus,

$$\Sigma p(1-p) = 0.4803$$

This gives:

$$S_{\log_{10} ED_{50}} = \sqrt{\frac{0.4803}{14}}$$

$$= 0.1852$$

The 95% confidence limits for the log ED_{50} are obtained from:

$$\log_{10}(95\% \ CL) = \log_{10}ED_{50} \pm 1.96 \ S_{\log_{10}} \ ED_{50} \quad \dots \dots \dots \text{Eq. 5.7}$$
$$= -4.967 \pm 1.96 \ (0.1852)$$
$$= -4.6040 \text{ and } -5.3300$$

which give reciprocal dilutions of challenge suspension of 1/214,000 and 1/40,200. Expressed in CFU, we have a lower 95% limit of 280 and an upper limit of 1500 for the LD_{50} of the suspension.

It can be seen from the above that the Spearman–Kärber procedure is simple to apply but suffers from the limitations that many sets of 'real-life' data will be unsuitable because of failure to meet the fairly restrictive criteria of (1) having equal-sized groups, (2) 0% and 100% responses at the ends of the series and (3) the 50% response somewhere near the middle. The reader who is familiar with animal work will probably have 'smelt a rat' in the set of figures taken to illustrate the procedure, because they are much more even and symmetrical than could reasonably be expected as routine in a virulence titration.

Perhaps a more serious criticism of Spearman–Kärber is its wastefulness of resources, in that a considerable proportion of the animals are set up to give responses far removed from 50% in order to have 0% and 100% responses, whereas logically one would think that since the ED_{50} is the quantity sought, it would be better to have the animals allocated to doses that would give responses in the range 20% to 80%. This leads us into a consideration of the probit procedure.

5.4.2 The probit method

This is a far more versatile and efficient way to estimate the ED_{50}, and its only significant drawback is the labour involved in the calculations when the method is applied in full. This, however, should not be a deterrent, because for routine use there are excellent computer programs (e.g. Larsson *et al.*, 1981) which make the task straightforward. These programs permit the evaluation of ED_{50} and confidence limits, not only on an individual sample being tested at several dilutions, but also the ED_{50}s of several preparations titrated in parallel.

The probit method differs from Spearman–Kärber in not requiring responses of 0% and 100%. Indeed, there is an actual preference for all the responses to be in the range of about 20% to 80%.

The probit method can be applied at different levels of sophistication and complexity. Here we shall just take it at two levels: the simple graphical use of probits and the simple arithmetic application to obtain approximate values of ED_{50} and 95% confidence limits. Further mention is made in Chapter 10, while Finney's (1962) monograph should be consulted for a comprehensive account.

5.4.2.1 Graphical approach

For comparative purposes, let us take the data from the virulence titration as used above and evaluate them by the probit method. First we shall use a graphical approach because of its simplicity. Fig. 5.4 shows the sigmoid curve obtained for the virulence titration data when we plot the proportion of mice killed against log-dilution of challenge suspension. This curve can be transformed into a straight line by expressing each proportion as a percentage and converting this in turn into a probit value by consulting Appendix A.6. The plot of probit of mortality against log-dilution of challenge suspension is given in Fig. 5.5. Note that we cannot plot probit values for 0/15 or 15/15 since such proportions are indeterminate on the probit scale. All we can do is to indicate that the points are above or below the line, as shown with arrowed points. Having fitted a straight line through the points by eye, we can read off the ED_{50} as that dilution corresponding to a probit of 5.0 (equivalent to 50%). From the graph as drawn, the ED_{50} is $10^{-4.8}$, which is quite close to the value of $10^{-4.967}$ as given by Spearman–Kärber.

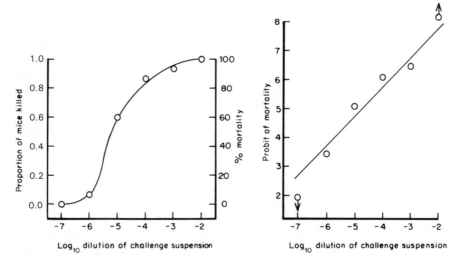

Fig. 5.4   The sigmoid dose–response curve for the virulence titration of a bacterial suspension

Fig. 5.5 Probit of the mortality versus log-dose of challenge organisms for the bacterial virulence titration. The straight line has been fitted by eye

5.4.4.2 Confidence limits of the ED_{50}

Having obtained an estimate of ED_{50} from Fig. 5.5, we can continue the calculations by obtaining estimates of the 95% confidence limits. For this, we need the standard error of the $\log_{10} ED_{50}$ ($S_{\log_{10} ED_{50}}$) which may be estimated approximately by:

$$S_{\log_{10} ED_{50}} = \frac{1}{b\sqrt{pn\bar{w}}} \quad \cdots\cdots\cdots\cdots\cdots \text{Eq. 5.8}$$

where

b = the slope of the log-dose/probit response line;
p = the number of dilutions in the titration = 6; this includes the two points which were not plottable on the graph;
n = the number of animals in each group = 15;
$\bar{w}$ = the average *weight* of the observations.

Let us take each of these in turn. b is defined as the increment in response per unit increment in $\log_{10}$ dose. If we take the dose values from 10^{-6} to 10^{-3}, i.e. 3 units on the $\log_{10}$ scale, we find an increment in probit response from 3.75 to 6.80. Note that these are read from the fitted line not from the tabulated probits plotted as points. The slope (b) therefore is $(6.80 - 3.75)/3 = 1.017$. p and n have already been defined. $\bar{w}$ is obtained from the right-hand column of Table 5.3.

Table 5.3 Virulence titration data set out for convenience of calculating observed and expected probits and their weights

Dilution of bacterial suspension	Mortality ratio	Mortality percent	Probit of observed mortality	Expected probit[a]	Weight (w) associated with expected probit[b]
10^{-7}	0/15	0	—	2.75	0.084
10^{-6}	1/15	6.7	3.50	3.75	0.353
10^{-5}	9/15	60	5.25	4.77	0.623
10^{-4}	13/15	86.7	6.11	5.77	0.512
10^{-3}	14/15	93.3	6.50	6.80	0.180
10^{-2}	15/15	100	—	7.80	0.025
					$\Sigma w = 1.777$
					$\bar{w} = 0.296$

[a]From the graph in Fig. 5.5
[b]Obtained by inserting each expected probit into the Table of Weighting Coefficients (Appendix A7).

Each observation, such as 9/15, has both an *observed* and an *expected* probit. The former is obtained by looking up the percentage equivalent of the proportion (e.g. 9/15 = 60%) in the probit table and finding the observed probit value. This provides the point for plotting on the probit versus log-dose graph. The *expected* probit, on the other hand, is obtained by extrapolating upwards from each dose value on the abscissa to intersect the fitted straight line. The point of intersection is then read across horizontally to the probit scale to get the expected probit. So unless the experimental point sits directly on the line, its observed and expected probit values will be different. Table 5.3 shows the observed and expected probits for each dose level. Note that it is possible to obtain *expected* probits for the 0 and 100% mortality results even although *observed* probits for these results could not be plotted. Thus these results are not discarded.

Now we come to the quantity referred to as the *weight*. The idea here is that doses which give a response close to 50% are of much more importance as determinants of the ED_{50} than doses giving responses close to 0% or 100%. The process of *weighting* the results is done quantitatively by means of a *weighting coefficient* associated with each expected probit value and is obtained from Appendix A7. The maximum value of the weighting coefficient is 0.6366 for a probit of 5.0. For probits greater or less than 5.0 this falls off symmetrically to values as low as 0.015 for probits of 2 or 8.

The right-hand column of Table 5.3 provides the weights for each expected probit. The sum of these values is $\Sigma w = 1.77$ and the mean weight, $\bar{w}$, is $\Sigma w/p = 0.296$.

We now have all the quantities needed to calculate $S_{\log_{10} ED_{50}}$ by Eq. 5.8:

$$S_{\log_{10} ED_{50}} = \frac{1}{1.017\sqrt{6 \times 15 \times 0.296}}$$

$$= 0.1905$$

which may be compared with the value 0.1852 from the Spearman–Kärber method.

Step 4. The 95% confidence limits may be established by Eq. 5.7:

$$\log_{10} (95\% \ CL) = \log_{10} ED_{50} \pm 1.96 \ (0.1905)$$
$$= -4.8 \pm 0.373$$
$$= -4.427 \text{ and } -5.173$$

Taking the antilogs and then reciprocals gives limits from 1/27,000 to 1/150,000 which are somewhat displaced from those given by the Spearman–Kärber method which were from 1/40,200 to 1/214,000. However, it must be emphasized that the above probit procedure is only approximate.

5.5 Estimation of Most Probable Numbers (MPN)

Bacterial viable counts are best done by colony-counting procedures where this is possible, and the statistical processing of such data is given in Chapter 6. There are, however, certain types of samples, and certain specific micro-organisms, where colony-counting is precluded or is technically difficult. For example, very muddy samples from rivers may be unsuitable for spreading on culture plates to give discrete colonies; and organisms with specific biochemical functions such as dentrification or ammonia-oxidation may not give character-istic and easily countable colonies, but may produce readily detectable chemical changes in liquid media. In water microbiology, before the advent of the membrane filter, the MPN method was the standard procedure for counting coliforms in water for human consumption. A microtechnique for MPN analysis which is claimed to be better than standard tube tests has been described by Rowe, Todd and Waide (1977).

5.5.1 Another instance of not doing your own barking

The principle of the MPN method is to make serial dilutions (usually 10-fold) of the sample and to inoculate constant volumes of each dilution into a fixed number of replicate tubes of liquid medium which are then incubated to allow growth of viable microorganisms. Any tube receiving one or more viable cells will exhibit growth, while the rest remain sterile. As an alternative, it may be more convenient to inoculate 10-fold decreasing volumes of the undiluted sample. The results of such a test might, for example, be 4/5, 2/5 and 0/5 for the proportions of positive tubes inoculated respectively with the sample undiluted, at 1/10 and at 1/100. What we want from such data is an estimate of the most probable number of bacteria per inoculum volume in the undiluted sample. The answer is 2.2, and the way we get it is to consult Appendix A9 which is due to Taylor (1962).

If we had used sets of 8 replicate tubes at each of our 10-fold dilutions, we could find the MPN in Appendix A10 and, with 10 replicate tubes per dilution, the corresponding table may be found in Norman and Kempe (1960) or Meynell and Meynell (1970).

The awkward microbiologist may, of course, wish to use other than 10-fold dilutions and number of replicate tubes other than 5, 8 and 10. In which case he or she will have to work out the MPN by calculation (see, for example, Russek and Colwell, 1983). Also, the earlier paper by Taylor (1962) should be consulted. I suspect, however, that most laboratory workers who find themselves using the MPN technique will prefer to set up their tests in one of the standard formats that allow the data to be evaluated directly from the tables.

5.5.2 Precision and confidence limits of MPN counts

Cochran (1950) has provided a set of numerical factors for the rapid and easy calculation of standard errors and 95% confidence limits for MPN counts. Assuming that the value for MPN is based on three 10-fold dilutions with 5, 8 or 10 replicate tubes at each, then the factors for the 95% confidence limits are as follows:

No. of replicate tubes per dilution	Factor for 95% confidence limits
5	3.30
8	2.57
10	2.32

Reproduced by permission of the Biometric Society.

To illustrate how these are used, let us take the previously cited example of three 10-fold dilutions giving 4/5, 2/5 and 0/5 positive tubes. The most probable count from the table is 2.2 viable organisms per volume inoculated. The 95% CL are obtained by dividing and multiplying the MPN value of 2.2 by the provided factor 3.30. This gives a lower 95% CL of 2.2/3.30 = 0.67 counts per inoculated volume (or 6.7 counts per 10 inoculated volumes). Likewise, the upper 95% CL is 2.2 × 3.3 = 7.26 counts per inoculated volume, or 72.6 per 10 volumes. From the width of these limits it is clear that the multiple tube method of determining a viable count is highly imprecise.

5.6   Is that Inheritance Mendelian? More χ^2

5.6.1   The human ABO blood-group system

In the human ABO blood-group system there are three major alleles, A, B and O, each of which is capable of occupying the ABO locus on each haploid chromosome. In each of our diploid cells we have 2 homologous loci, so that the possible genotypes are AA, AO, AB, BB, BO and OO. Since the O gene is 'silent' and the A and B genes are codominant, the corresponding blood-group phenotypes are A, AB, B and O. In former years there was much controversy about this scheme and Wiener (1943), among others, made extensive surveys to see if supporting data could be obtained. We shall take just a single example in which he determined the blood groups of 129 children who were the offspring of marriages in which both parents were of blood group AB. The observed numbers in the children were:

Group A	28
Group B	36
Group AB	65
Total	129

According to the theory discussed, the predicted distribution of blood groups in offspring from AB × AB matings is:

Group A	¼
Group B	¼
Group AB	½

Therefore in a population of 129 individuals, the expected numbers would be:

Group A	32.25
Group B	32.25
Group AB	64.5

One can see without doing a χ^2-test that the observed and expected numbers are in close agreement. Nevertheless, to confirm this impression we apply Eq. 5.3:

$$\chi^2 = \frac{(28-32.25)^2}{32.25} + \frac{(36-32.25)^2}{32.25} + \frac{(65-64.5)^2}{64.5}$$

$$= 1.00$$

With 3 groups contributing to the total of 129 observations, the d.f. $= 2$, since when the numbers in 2 of the groups have been fixed independently, the third group no longer has freedom to vary. For 2 d.f. the 5% point of χ^2 is 5.99 (Appendix A4) and there is thus no significant difference between the observed and expected numbers.

5.6.2 The genetics of serum complement in mice

In a study of the inheritance of a defect in serum complement in inbred mice, Cinader, Dubiski and Wardlaw (1966) made hybrids and back-crosses from parental strains which either possessed or lacked a functional complement system in the serum. The genotype of a complement-containing strain may be designated C.C. and of a deficient strain as c.c., so that the F_1 hybrid is C.c. Such hybrids possessed complement, showing that C is dominant over c. When back-crosses of C.c. $\times$ c.c. were made with two of the strains, 85/188 offspring were complement-containing and 103/188 were complement-lacking. The question of interest for this chapter is whether these two *observed* ratios depart significantly from the *expected* ratios of 94/188 and 94/188 — that is, ratios expected on the assumption of unifactorial dominant inheritance.

This is a simple exercise in χ^2 where we apply Eq. 5.3 to the above observed and expected numbers:

$$\chi^2 = \frac{(85-94)^2}{94} + \frac{(103-94)^2}{94}$$

$$= \quad 1.72$$

Here the degrees of freedom (d.f.) $= 1$, because with only 2 categories of phenotype, once we know the number in one of these, the other is no longer independent with a given total number of animals.

Since the 5% point of χ^2 with 1 d.f. is 3.84, a found $\chi^2 = 1.72$ is not significant, and the ratios 85/188 and 103/188 do not depart from a 1:1 ratio by more than can be explained by random-sampling fluctuations. The hypothesis of unifactorial dominant inheritance of the complement component in the mice is therefore not challenged.

5.7 Goodness-of-Fit to a Normal Distribution

In Chapter 2 we showed by graphical methods and with the aid of probits (Fig. 2.10, Section 2.6.2) that the 135 pipetting results gathered by a class of

Table 5.4 Protocol for calculating χ^2 from the already grouped 135 pipetting results to check for departures from normality

Upper boundary (UB) in working mg (1)	Actual no. of results (O) (2)	Expected probit (EP) (3)	% corresponding to EP (4)	Expected accumulated no. of results (5)	Expected no. of results in group (E) (6)	$\dfrac{(O-E)^2}{E}$ (7)
7	4	3.096	2.8	3.78	3.78	0.013
14	8	3.475	6.4	8.64	4.86	2.029
21	8	3.854	12.6	17.01	8.37	0.016
28	11	4.232	22.1	29.83	12.82	0.258
35	12	4.611	34.9	47.11	17.28	1.613
42	16	4.990	49.6	66.96	19.85	0.747
49	29	5.369	64.4	86.94	19.98	4.072
56	19	5.748	77.3	104.35	17.41	0.145
63	14	6.127	87.0	117.45	13.1	0.062
70	4	6.506	93.4	126.09	8.64	2.492
77	6	6.884	97.1	131.08	4.99	0.204
84	2 } 4	7.263	98.8	135.38	2.3 } 3.38	0.114
91	2	7.642	99.6	134.46	1.08	

$$\Sigma = 11.765$$

$N = 135$
$\mu = 42.1815$
$\sigma = 18.477$

$$EP = \left(\frac{UB - \mu}{\sigma} \right) + 5$$

$k = 12$
$\text{d.f.} = 9$
$p > 10\%$

students were quite a good fit to a normal distribution. Now we can use the χ^2-test to investigate the goodness-of-fit by a more objective procedure to replace naked-eye judgement. Our calculations start with the 135 results already grouped into 7-mg-wide classes, as shown in Table 5.4, the first two columns of which are carried forward from Chapter 2 (Table 2.4). This is an exercise in the analysis of proportion data since the number of observations in each histogram bar can be expressed as a whole number divided by 135.

The protocol for calculating χ^2 is set out in Table 5.4 to which the following notes apply:

Step 1. In columns 1 and 2 are set down the upper boundary (*UB*) of each class interval used for grouping the data, and the number of observations, both having previously been recorded in Table 2.4.

Step 2. Using the grouping procedure (Section 2.5.2) and Eqs 2.2 and 2.3, we have already calculated $\bar{y} = 42.1815$ and $s = 18.477$ for the 135 observations. Because $N = 135$ is so large, we can reasonably take these values of $\bar{y}$ and s as close approximations to the underlying μ and σ, respectively.

Step 3. For each value of *UB*, calculate an *expected probit* (*EP*) from the formula:

$$EP = \frac{UB - \mu}{\sigma} + 5 \dots\dots\dots\dots\dots\dots\dots\dots Eq.\ 5.9$$

For example, putting $UB = 7$ gives

$$EP = \frac{7 - 42.1815}{18.477} + 5$$

$$= -1.904 + 5$$

$$= \quad 3.096$$

which gives the first entry in column 3. The figures in this column are thus the probits (corresponding to each *UB*-value) that lie directly on the fitted straight line of Fig. 2.10. Note that expected probits and observed probits would be numerically the same if we were dealing with an infinitely large sample from a perfectly normal distribution.

Step 4. Next consult the table of probits (Appendix A6), and for each probit value in column 3 of Table 5.4 read off the corresponding percentage. This is the reverse of our usual procedure with a probit table where we read off the probit corresponding to a given percentage. We find that for probit $= 3.096$ the corresponding $\% = 2.8$ (approximately). We enter this and the other $\%$ values in column 4.

Step 5. Multiply each column 4 value by 1.35. This gives us the expected accumulated number of results from a population of size $N = 135$, since the $\%$ scale of column 4 corresponds to $N = 100$. Thus, the first entry, 2.8, in column 4 becomes 3.78 in column 5. This is the theoretical number of observations (making the usual allowances for the fact that for theoretical purposes you do not need to have a whole number) which would have been accumulated between 0 and 7 mg in a sample of size $N = 135$ taken from a perfect normal distribution of $\mu = 42.1815$ and $\sigma = 18.477$ and with no sampling fluctuations.

Step 6. Convert the 'expected *accumulated* number of results' in column 5 to 'expected number of results' in column 6 by successively subtracting from each column 5 entry the number above it. For this purpose the number above the first entry in column 5 is taken as zero which, subtracted from 3.78, leaves 3.78, the first result for column 6. Similarly, $8.64 - 3.78 = 4.86$, the second entry for column 6, and so on. These values in column 6 are the *expected numbers* of results within each 7-mg-wide class, with which the *observed numbers* (O) of column 2 may be compared.

Step 7. Note that at the extreme ends of the distribution in column 6 we have some small expected numbers of < 5. The rule here for χ^2 purposes is that we are allowed to have the 2 extreme E-values as low as about 1, provided that the other E-values are not much less than about 5. We should, therefore, lump together the two last values of column 6 as shown, the rest being taken as not violating the rule.

Step 8. Calculate $(O - E)^2/E$ by taking each O-value from column 2 and the corresponding E-value from column 6 and entering the result in column 7. The sum of the column 7 values then gives χ^2 by the formula:

$$\chi^2 = \Sigma(O - E)^2/E \quad \dots\dots\dots\dots\dots\dots\dots\dots\text{Eq. 5.3}$$

$$\chi^2 = 11.765$$

The rule for d.f. in this calculation is that if χ^2 is based on k groups, the d.f. $= k - 3$. Here our χ^2 is based on $k = 12$ groups, therefore d.f. $= 9$. Consulting the χ^2 table (Table A4) for 9 d.f. shows that our *found* value of χ^2 corresponds to P greater than 10%. Thus, the null hypothesis which states that 'our pipetting data do not depart from a normal distribution by more than can reasonably be explained by random-sampling fluctuations' is not rejected. We may therefore conclude that the departures from normality are 'not significant'.

Referring back to Section 2.4.3 it was stated that the figures 76 and 59 for the numbers of pipetting results above and below the mean do not depart significantly from the expected value of 67.5 and that a χ^2-test could be done to establish this.

Taking Eq. 5.3 with $E = 67.5$ and $O = 76$ and 59, we get:

$$\chi^2 = \frac{(76 - 67.5)^2}{67.5} + \frac{(59 - 67.5)^2}{67.5}$$

$$= \frac{8.5^2 + 8.5^2}{67.5}$$

$$= 2.14$$

This *found* value of χ^2 does not reach the tabulated (Table A4) 5% point at 3.84 for 1 d.f. Therefore the null hypothesis stands, and we conclude that 'the difference in distribution of pipette deliveries above and below the mean value $\mu = 42.1815$ is explainable by random-sampling fluctuations'.

It should be noted that we can never *prove data to be* exactly normally distributed—in fact is it highly improbable that experimental data ever are *exactly* normally distributed. All we can do is to find out if departures from normality are serious or trivial. It is rather like in chemistry where one can never get a substance in a state of absolute purity—in the sense that there is not a single atom of extraneous material contained in it. Nevertheless, one can work perfectly well with materials whose content of impurity is insufficient to be troublesome. So it is in statistics.

To pursue the chemical analogy, the smaller the level of 'impurity' in the normality of our distribution, the larger the sample we would need to detect it. Suppose, for example, in the pipetting results under discussion we had 1,350 observations instead of 135. Suppose also, for simplicity, that the distribution of the observations was exactly the same as we have had here except that there were 10 times as many in each 7-mg-wide class. This would give us $\chi^2 = 117.65$ instead of 11.765, and this would be highly significant, showing that *such* a pattern of 1,350 observations indeed departed significantly from normality. On the other hand, if we were *actually to go ahead and collect* the 1,350 observations by, say, getting 10 classes of 15 students each to collect 9 observations per student, we might well find this very large assemblage of data to approximate well to a normal distribution.

With the 135 actual observations that we have, all we can say is that the apparent departures from normality are within the limits that could easily arise through random-sampling fluctuations. Also, the specific checks for skewness and kurtosis when done by the procedures given by Snedecor and Cochran (1967, p. 86), show that neither of these two 'specific impurities' is present at significant levels.

6

How to Deal with Count Data

Truth is rarely pure, and never simple.

Oscar Wilde (1854–1900)

6.1 From Bacterial Colonies to Umbrellas Left on Buses

Contrary to what some readers may think, statisticians are not necessarily lacking in wit, imagination or the common touch, and we conveniently start this chapter with a collection of items to illustrate this point. We are going to be concerned mainly with the Poisson distribution and its application to count data, such as bacterial colony counts and radioactivity counts. But before we come to these, let us consider some non-laboratory examples of count data to which the Poisson distribution has been reported to apply:

- the number of umbrellas left per day on buses in a large city;
- the number of flying-bomb hits per square kilometer of London during the Second World War;
- the annual number of men killed by kicks from horses in the Prussian army in the days before mechanized transport;
- the number of death notices per day for men over 85 in the obituary columns of the London *Times*.

What is the factor that is common to each of these and to bacterial colonies and radioactive disintegrations?

(1) The data consist of *whole-number observations*, i.e. *counts*. You cannot have 4.61 bacterial colonies on a culture plate, or 4.61 umbrellas left last Tuesday on Glasgow city buses.

(2) Each particle or event or object that is recorded is a very small fraction of the total number 'at risk', i.e. the bacterial colonies actually on the plate originate from a very small fraction of the total number of organisms present in the bulk material (air, water, broth culture, mouse blood, etc.) from which the sample was taken. Likewise, the umbrellas left on the buses are a very small fraction of the total number being carried on buses on that day.

(3) The particles or objects counted are *independent* of each other and therefore occur randomly, i.e. the leaving of an umbrella on one bus by one passenger in no way influences the leaving of other umbrellas on other buses by other passengers.

(4) The number of particles or events counted in each unit of time or space is usually fairly small. (This is true of radioactivity counts if the time interval is made sufficiently short, and of bacterial colony counts if the dilution is sufficiently great.)

We can now generalize by stating that the Poisson distribution, or Law of Small Probabilities, is a concept that applies to the counts of: Particles, objects or events that are randomly distributed in space or time, and where a large population of particles or events are potentially available, i.e. 'at risk', but only small numbers are 'captured' within the span of each unit of time or space that is sampled or observed.

There are four main things to say about the Poisson distribution:

(1) it describes, at least approximately, many types of count data generated by the experimental biologist: bacterial colony counts, virus plaque counts, total cell counts and radioactivity counts;

(2) it provides a theoretical basis for analysing a variety of one-hit phenomena such as phage-bacteria and radiation/genome interactions;

(3) if the count per unit of volume or time is sufficiently large, the Poisson distribution approximates to a normal distribution;

(4) Living organisms in their *natural environments* are frequently *not* Poissonly distributed, i.e. they are *not* distributed at random, but tend to show either clustering, aggregating and herding or, the opposite, i.e. undue spacing out (territorialism).

The following sections will deal with each of the above features in turn.

6.2 Total Cell Counts and the Poisson Distribution

6.2.1 The Helber slide counting chamber

Before the widespread availability of the Coulter Counter, it was commonplace to do total cell counts on suspensions of bacteria or of cells from other sources, such as blood, with the Helber, or similar (Thoma, Neubauer, Petroff–Hausser) counting chambers which consist of special, ruled microscope slides. These are still useful for special purposes or where a Coulter Counter is not available. For the present discussion they provide a convenient source of data for introducing the Poisson distribution.

In the Helber type of slide chamber there is a central, sunken platform with rectangular grid rulings covering an area of 1×1 mm and separated from the overlying coverglass by a depth of 0.02 mm. The square-ruled area has a pattern of 25 large squares, each of which is divided into 16 small squares. The side of a small square is 0.05 mm and the volume above each small square is 0.5×10^{-7} cm^3. The normal procedure with the counting chamber is to fill it with a cell suspension whose concentration, based on a preliminary count, yields an average of between 1 and 5 cells per small square. You then count the cells in a total of 80 small squares (5 fields) and obtain the average count per small square. This value multiplied by 2×10^7 (the reciprocal of 0.5×10^{-7}) gives the count per cubic centimetre. Because of manipulative errors in placing the coverslip properly, full reliance should not be put on the counts from a single filling of the chamber. Instead, the chamber should be rinsed, dried and refilled, and the counts on 4 independent fillings obtained. Let us now take a specific numerical example.

6.2.2 An example of Helber slide data

An overnight nutrient broth culture of *Escherichia coli* NCTC 8623 was diluted 1:20 in phosphate-buffered formol-saline containing a trace of Teepol (see Meynell and Meynell, 1970, p. 20) and counted by the above procedure to yield the results set out in Table 6.1. It will be noted that 4 independent fillings of the chamber were made and that a total of 320 small squares contained 923 bacterial cells. This gives the average count per small square as 2.88, and the count in the *undiluted* broth culture as:

$$2.88 \times 2 \times 10^7 \times 20 = 1.15 \times 10^9 \text{ bacteria per cm}^3.$$

So much for the arithmetic, now for some statistics.

Table 6.1   Example of counts of E. *coli* cells in the Helber chamber

Chamber filling no.	No. of small squares counted	No. of E. *coli* cells	No. of cells per small square
1	80	210	2.62
2	80	268	3.35
3	80	244	3.05
4	80	201	2.51
Total	320	923 Average	2.88

6.2.3   Analysis of competence in filling the chamber reproducibly

The main technical difficulty in using the Helber chamber is in putting the cover glass on properly at each filling, since any inconsistency here affects the depth, and hence the volume, of liquid above the squares. The results in Table 6.1 show a range in average count per small square from 2.51 to 3.35, a difference of 25%. Is this difference due to inconsistencies in filling the chamber, or is it attributable to random-sampling fluctuations in the counts themselves? A χ^2-test will answer this question.

Step 1. Set up a null hypothesis which states that 'the difference in the total number of cells counted at each filling of the chamber is no greater than can be explained by the random-sampling fluctuations in making the counts'.

Step 2. Calculate the expected (E) count per filling as the average of the 4 totals, i.e.

$$E = \frac{210 + 268 + 244 + 201}{4} = \frac{923}{4} = 230.75$$

Step 3. Taking the total count from each filling as an observed (O) value, calculate χ^2 from:

$$\chi^2 = \Sigma(O - E)^2/E \ldots\ldots\ldots\ldots\ldots\ldots\ldots\ldots\ldots\ldots\ldots\ldots\ldots\ldots\text{Eq. 6.1}$$
$$\text{(and Eq. 5.3)}$$

$$\therefore \chi^2 = \frac{(210 - 230.75)^2 + (268 - 230.75)^2 + (244 - 230.75)^2 + (201 - 230.75)^2}{230.75}$$

$$= 2878.75/230.75$$
$$= 12.48$$

With $k = 4$ totals, we look up (Appendix A4) the tabulated value of χ^2 for $k - 1 = 3$ degrees of freedom and find that 12.48 corresponds to a probability of between 1% and 0.1%.

Step 4. The null hypothesis is therefore rejected and we can declare that there *is* a highly significant heterogeneity between individual fillings of the chamber.

Technically, the only fully satisfactory way to get around this difficulty in the placement of the cover glass is, at each filling, to measure the actual depth of liquid in the chamber by the interferometric device of Norris and Powell (1961). However, this equipment is not commonly available.

Table 6.2   Examples of the distribution of *E. coli* cells among the squares of the Helber slide

Field 1

2	3	4	2
5	0	1	3
1	2	4	2
4	3	1	7

Field 2

6	1	5	1
2	3	4	2
4	1	2	3
2	3	2	1

Field 3

6	2	3	2
1	4	2	4
2	3	3	1
2	1	1	5

Field 4

2	2	1	4
4	3	2	3
2	3	5	3
1	2	1	0

Field 5

2	4	0	1
5	1	3	2
4	2	0	2
4	3	1	1

6.2.4   Studying how the bacteria are distributed over the squares

In the ordinary usage of the Helber slide chamber, one simply counts the total number of cells over 80 small squares, but such data do not permit investigation of the postulated underlying Poisson distribution. To get data for this, we have to take the extra trouble of recording the number of cells over each individual square. This is certainly worth doing periodically to check that the distribution is Poisson even though it would be too time-consuming to do routinely with every filling of the chamber. Table 6.2 provides an example of the distribution of the individual *E. coli* cells in each small square from the 5 fields examined. This is, in fact, a detailed break-down of the total count of 201 recorded for filling no. 4 in Table 6.1. Note that in making these counts it is important to record accurately all cells that lie directly on the grid lines, but also to avoid counting them twice. A useful convention with any particular square is to count cells on the top and right-hand lines as lying within that square, but to avoid counting cells on the bottom or left-hand lines. The latter cells will be counted as contained in adjacent squares.

Our first step in processing the array of counts in Table 6.2 is to reduce their bulk by summarizing them as a frequency-distribution table. This has been done in Table 6.3 which contains a wide column 2 for accumulating as slashes, the

Table 6.3   Frequency-distribution relationship produced from the array of results in Table 6.2

No. of bacterial cells per square	No. of squares with chosen number of bacterial cells	Total number of squares	No. of squares as a proportion of 1.0
0	////	4	0.05
1	/HI /HI /HI ///	18	0.225
2	/HI /HI /HI /HI ///	23	0.2875
3	/HI /HI /HI	15	0.1875
4	/HI /HI //	12	0.15
5	/HI	5	0.0625
6	//	2	0.025
7	/	1	0.0125
	Total	80	1.0000

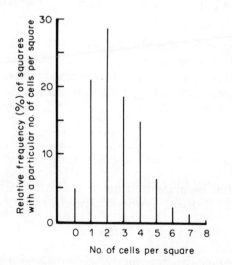

Fig. 6.1   Frequency-distribution diagram of the data in Tables 6.2 and 6.3

tally of squares with each occupancy number. In columns 3 and 4 we total the number of slashes for each occupancy number and then express these totals as a fraction of 80, the overall number of squares that were counted.

Fig. 6.1 plots the corresponding frequency-distribution diagram with the ordinate labelled in per cent. This diagram bears some resemblance to a normal-distribution histogram, but also differs in that the individual frequencies are represented as separated lines rather than as connected bars. This is because the distribution is *inherently discontinuous*, reflecting the count data on which it is based.

Fig. 6.1 thus presents an *observed* frequency-distribution of the counts per square. The next point of interest is to see how closely this conforms to a Poisson distribution.

6.2.5 The Poisson distribution

The Poisson distribution, due to the French mathematician S. D. Poisson, is described by the equation:

$$P = \frac{m^x \cdot e^{-m}}{x!} \qquad \dots\dots\dots\dots\dots\dots\dots\dots\dots \text{Eq. 6.2}$$

where

P = the probability, or frequency, of occurrence of squares containing a particular number of cells; P is expressed as a decimal of 1.0;

m = the *average* number of cells per square; this does not need to be a whole number and usually will not be;

x = the occupancy number of a square; x can have only whole-number values 0, 1, 2, 3, 4, ...

With the example under discussion, we have 201 bacteria distributed over 80 squares, giving $m = 201/80 = 2.5125$.

We can then use the Poisson formula to calculate the *expected* number, or proportion, of squares with $x = 0, 1, 2, 3$, etc. bacteria, if the average per square is 2.5125. For example, if we put $x = 0$ in Eq. 6.2, we get:

$$P = \frac{(2.5125)^0 \cdot e^{-2.5125}}{0!}$$

$e^{-2.5125}$ may readily be obtained from a calculator with an e^x button as 0.081065. Then $(2.5125)^0 = 1$, as does 0! Therefore, for $x = 0$:

$$P = 0.081065$$

This is our *expected* relative frequency-of-occurrence of empty ($x = 0$) squares, and may be compared with the *observed* value (Table 6.3) of 0.05. Expressed as the actual number of squares out of 80, the expected figure is $80 \times 0.081065 = 6.48$, which may be compared with the found value of 4. Note that there is no logical inconsistency in the expected number not being an integer, since the expectation is that for an infinitely large number of squares, not just a sample of 80 squares.

If we now put $x = 1$ into Eq. 6.2, we get:

$$P = \frac{(2.5125)^1 \cdot e^{-2.5125}}{1!} = 0.2037,$$

which should be compared with the observed value of 0.225 (Table 6.3); or $80 \times 0.2037 = 16.3$ squares (expected) may be compared with 18 squares (observed).

Similarly, by successively substituting $x = 2, 3, 4, \ldots$ into Eq. 6.2 we can obtain the other expected P-values for the distribution. These calculations are easy to do with a calculator with buttons for y^x, e^x and $x!$ and two memories in which 2.5125 and $e^{-2.5125}$ can be stored. Column 3 of Table 6.4 lists all of the expected frequencies calculated in this way, while column 4 gives the expected number of squares in a random sample of 80. A quick visual comparison of the expected (E) numbers in column 4 with the observed numbers (O) in column 2 indicates that our $E.\ coli$ count data conform rather closely to those expected from a Poisson distribution. This impression is confirmed by completing the calculation of χ^2 by Eq. 6.1.

Table 6.4 Calculation of χ^2 to test the goodness-of-fit of the observed occupancy rates in the Helber chamber to a Poisson distribution

Occupancy no. (x)	Observed no. of squares (O)	Expected frequency (P)[a]	Expected no. of squares (E)[b]	$\frac{(O-E)^2}{E}$
0	4	0.0811	6.49	0.955
1	18	0.2037	16.30	0.177
2	23	0.2559	20.47	0.313
3	15	0.2143	17.14	0.267
4	12	0.1346	10.79	0.136
5	5	0.0676	5.41	0.031
6	2 ⎫	0.0283	2.27 ⎫	⎫
7	1 ⎬ 3	0.0100	0.81 ⎬ 3.15	⎬ 0.007
>7	0 ⎭	0.0045[c]	0.07 ⎭	⎭
Total	80	1.0000	79.75 (should be 80.0)	1.89

[a]By substituting values of $x = 0, 1, 2, 3, \ldots$ into Eq. 6.2, as described in the text.
[b]Simply $P \times 80$.
[c]Obtained by adding up all the other P-values and subtracting them from 1.0000.

Note that the data for the last 3 categories of x have been pooled so as to avoid having very small values of E (see Section 5.7). This leaves $k = 7$ values of $(O-E)^2/E$ and $k-2 = 5$ degrees of freedom (d.f.) for χ^2. The rule for counting the number of d.f. for χ^2 in this context is:

d.f. = (no. of categories or classes) – (no. of estimated parameters) – 1.

The estimated parameter is $m = 2.5125$ used in the Poisson formula.

For d.f. = 5, the found value of $\chi^2 = 1.89$ is not significant (5% point = 11.07), and therefore the observed occupancy rates of the squares do not differ from those expected from the Poisson formula by more than that due to random-sampling fluctuations.

6.2.6 The changeable shape of the Poisson distribution

We have already seen, in Fig. 6.1, the approximate shape of the Poisson distribution for $m = 2.5125$. Let us now see how the shape varies when we substitute lower and higher values of m into Eq. 6.1. Fig. 6.2 shows the Poisson distributions for $m = 0.1$, 1.0 and 10.0. With $m = 0.1$, the majority of squares (if we continue to think of the Helber chamber) are empty and so we have a very large bar at $x = 0$, a small bar at $x = 1$ and essentially nothing else, because of the very low probability (0.0047) of having a square with 2 or more bacteria when the suspension is so dilute.

The distribution where $m = 1.0$ is interesting because the first 2 bars are exactly the same height at $P = 0.3679$. We shall return to this when we discuss one-hit phenomena in Section 6.4.1. For the moment, however, note how the distribution is moving towards symmetry as m gets larger.

At $m = 10.0$ the Poisson distribution has the outline of an almost symmetrical bell-shaped curve although, of course, it is still an array of vertical bars and not a continuous line. Here we come to an important fact: as m becomes 'large', say > 20, the outline of the Poisson distribution approximates very closely to the smooth curve of a normal distribution. But it is a normal distribution with a very special 'degree of spreadoutness', in that the standard deviation (σ) is equal to the square root of the mean (μ). Even with $m = 10$, the approximation of the Poisson to the normal is very close, as can be seen from Fig. 6.2 where, superimposed on the Poisson plot, is the curve of the normal distribution for $\mu = 10$ and $\sigma = \sqrt{10}$. There is still, at $m = 10$, a small degree of asymmetry in the Poisson distribution but, aside from that, the approximation is very close. Therefore we would not be very far wrong by taking $m = 10$ rather than $m = > 20$ as being sufficiently 'large' in the context of this approximation.

This convenient approximation of the Poisson distribution to a normal distribution has useful consequences for calculating the standard error of the mean (SEM) and confidence limits of count data (next section). It also has some *undesirable* consequences for analysis of variance (see Section 6.5.5).

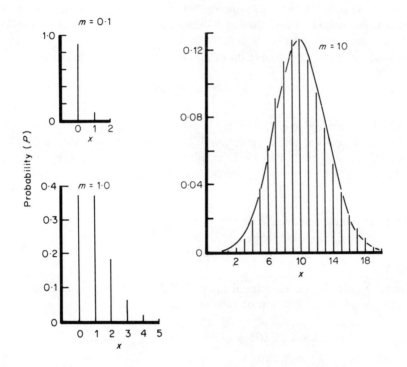

Fig. 6.2   The changeable shape of the Poisson distribution as m increases from 0.1 to 1.0 and to 10. The diagram for $m = 10$ has superimposed on it the graph of the normal distribution for $\mu = 10$ and $\sigma = \sqrt{10}$

6.3   Standard Error and Confidence Limits of Counts

6.3.1   Helber chamber data

Referring back to Table 6.1 where some representative Helber chamber counts were reported, we had a total of 923 bacterial cells counted over 320 squares, giving an average of 2.88 cells per square. As with most other averages that we calculate, we should not be content merely to give the average but should attach an indication of precision, with the standard error of the mean, or we should give an *interval estimate* in the form of 95% or 99% confidence limits. Whichever we do—and here we shall do both—the important idea to grasp is that the figure that we take for the calculation is the *total actual count* (of 923 in this example) and *not* the average count of 2.88 cells per square.

This is the crux of the argument: if the cells are Poissonly distributed in the counting chamber, then the size of square over which they are counted is quite arbitrary. We can therefore regard our count of 923 cells as having been made over one very large square whose area is equal to 320 small squares. We can therefore take this single count of 923 as an estimate of m, the average count over the hypothetical very large square. A Poisson distribution with such a large value of m should be a very close approximation indeed to a normal distribution with $\mu = m$ and $\sigma = \sqrt{m}$.

Thus, we immediately have an estimate of standard deviation as:

$$s = \sqrt{923} = 30.3809.$$

The SEM is equal to $s/\sqrt{N}$ and in this case $N = 1$ because we are treating our count of 923 cells as *one* observation. The standard error of the mean (SEM) is also therefore equal to 30.3809. At this stage we can, if we wish, express the count and its SEM in terms of the small square by dividing by 320, so that we get:

$$\text{count per small square} \pm \text{SEM} = \frac{923}{320} \pm \frac{30.3809}{320}$$

$$= 2.884 \pm 0.095$$

Since the volume over each small square is $0.5 \times 10^{-7} \, \text{cm}^3$, we can express the count $\pm$ SEM in the dilution of cells used as:

$$2 \times 10^7 \, (2.884 \pm 0.095) = (5.77 \pm 0.19) \times 10^7 \text{ per cm}^3$$

The 95% confidence limits are obtained by multiplying the SEM by 1.96 and then subtracting and adding the result from the mean. This gives us:

$$95\% \text{ CL} = (5.77 \pm 1.96 \times 0.19) \times 10^7$$
$$= 5.40 \times 10^7 \text{ and } 6.14 \times 10^7 \text{ per cm}^3$$

If we want the 99% confidence limits, the SEM should be multiplied by 2.576 instead of 1.96. With the present data this gives us:

$$99\% \text{ CL} = 5.28 \times 10^7 \text{ and } 6.26 \times 10^7 \text{ per cm}^3.$$

The essence of the above discussion is to recognize and accept that the precision of a count — any count — is determined by the total number of objects or events that are counted and not by the average number of objects per square (or per other unit of volume or time). The maximum practicable precision with a counting chamber is obtained by counting 500–1,000 organisms, since any appreciable increase is achieved only by counting many thousands. Thus, if we express the precision as the ratio of the SEM/total count, then for a count of 500, the ratio is $\sqrt{500}/500 = 0.0447$, or about 5%. If you do twice as much work and count 1,000 cells, the precision is reduced only to $\sqrt{1,000}/1,000 = 0.0316$, or about 3%. To get the ratio down to 1% requires a count of 10,000, which with a Helber chamber would indeed be a labour of love. Also, to push the precision to such an extent would be unrealistic in view of the technical difficulty in filling the chamber reproducibly.

6.3.2 Coulter Counter and radioactivity counts

From a statistical standpoint, the outputs from a Coulter Counter and from a radioactivity beta or gamma counter can be considered together, at least for elementary discussions. In both cases the count data are assumed to follow a Poisson distribution and, by taking sufficient counts, we can achieve any level of precision desired. There are practical limits, however, since the total error of count data is made up of two independent components: the error in preparing the sample for counting, i.e. the diluting and dispensing errors in delivering the sample to the machine, and the Poisson error in the actual counting. Let us suppose that the diluting and dispensing of the sample is done with equipment that is volumetrically accurate to, say, $\pm 2\%$, and that there are, say, three dispensing manipulations with the various reagents in the system, so that the sample preparation error could be as high as $\pm 6\%$. With such an error built into the delivery of the samples to the machine it would seem scarcely worthwhile to do the counting to better than $\pm 1\%$.

If by '$\pm 1\%$' we mean an SEM that is $\pm 1\%$ of the count, then this would be achieved by 10,000 counts, since $\sqrt{10,000} = 100$, which is 1% of 10,000.

If, however, by '$\pm 1\%$' we mean 95% CL of $\pm 1\%$, then the number of counts would have to be increased by 1.96^2, or nearly 4-fold, to 38,400.

Moving from the introductory and theoretical to the practical and routine aspects, the basic differences between the Coulter Counter and a beta or gamma counter are (a) the automation of the latter and (b) the correction for coincidence in the former at high counts. The Coulter Counter is normally fed with the cell suspensions individually, and in a given type of experiment or routine assay, the various samples are probably not going to vary by much more than about 10-fold. Indeed, if they do vary by much more than this one would tend to use different dilutions so as to make the counts from the machine more uniform.

With a radioactivity counter the position is somewhat different in that the machine can be loaded with a large number of samples which are then fed and counted automatically either for a pre-set time or for a pre-set number of counts. The problem with the former setting is that low-activity samples may not be counted for long enough to give the desired level of precision, while the problem with the latter is that the low-activity samples may take an inordinately long time. If, as is usually the case, large numbers of samples have to be counted automatically, and if one wishes to optimize the use of an expensive piece of laboratory equipment, a sophisticated approach is required. Such an approach also has to take account of the errors involved in preparing the samples for the machine, since there is no point in counting to $\pm 0.1\%$ if there may be a 5% error in sample preparation and delivery.

The widespread development of radioimmuno assays in recent years has led to a high level of sophistication in the computer programming of radioactivity counting equipment. An important concept to emerge is that of 'optimized precision counting' which takes note of sample-preparation error and balances it against counting error, so that an optimum counting scheme can be adopted for maximum throughput and cost efficiency. This is a considerable advance on simply switching the counter to a pre-set time or pre-set number of counts, and should be pursued in detail by those who use radioimmuno assays routinely (see Rodbard, 1974).

6.3.3 Colony and plaque counts

We can consider bacterial colonies on an agar plate and phage plaques on a bacterial lawn, or animal virus plaques on a tissue culture monolayer, as conceptually the same for the present discussion. Thus, a single bacterial culture plate with y colonies on it is physically equivalent to the 80 squares of the Helber chamber with y cells counted at a given filling. Provided that the number of colonies is more than about 20, we can take the SEM as equal to the square root of the count and the 95% CL as $y \pm 1.96$ SEM. Thus, if we have 32 colonies on a plate, the 95% CL are approximately $32 \pm 1.96\sqrt{32}$, which is 20.9 and 43.1. These limits are only approximate because the Poisson is not *fully* normalized at such relatively low counts. The exact 95% CL can be obtained from Appendix A8 and will be found to be slightly different, i.e. 21.888 and 45.175.

In making colony counts we may wish to set up replicate plates and then check that any variation between replicates is within the limits of expected counting fluctuations. If they are not, then significant technical error is probably creeping into the preparation and dispensing of the samples. Suppose, for example, we have 2 replicate plates with 100 and 130 colonies. Is this difference within the limits of expected counting error? We can do a χ^2-test in which we regard the counts of 100 and 130 as estimates of an underlying ('expected') count of 115. We then have

$$\chi^2 = (100 - 115)^2/115 + (130 - 115)^2/115$$
$$= 3.91$$

This is just significant at the $P = 5\%$ level for 1 degree of freedom (d.f.) (tabulated value = 3.84). We could therefore conclude that counts of 100 and 130 show significantly greater variation than can reasonably be explained by random-sampling fluctuations and we should therefore suspect that there have been volumetric inconsistencies in delivering the test samples to the plates.

6.4 Beyond Simple Counting

6.4.1. Investigation of one-hit and allied phenomena

There are various phenomena which may be described as 'one-hit' because they depend on the delivery of one effective unit of an agent to a sensitive target site, whereupon there is an observable response. Examples are the formation of a colony on an agar plate, the initiation of bacterial growth in a tube of broth, the alteration of a gene by a quantum of radiation, the infection of a bacterium by phage and the induction of certain infections in animals where one viable microorganism is sufficient. The Poisson distribution provides a theoretical base for each of these, with the sensitive target site being conceptually equivalent to a grid square in the Helber chamber.

In its simplest form, the one-hit idea envisages that the target sites are homogeneous in size and in sensitivity, and that the agents being delivered to them are (a) independent of one another and (b) equally effective in inducing the one-hit change, whatever it may be. If the system meets these requirements, then there is a primary interest in determining the dose, or concentration, of the agent that produces a change in 63% of the available target sites, as will now be explained. Let us take as an example what would be expected to happen if tubes of broth were inoculated with 1.0-ml amounts of a bacterial suspension containing 1 cfu per ml. At a first guess one might think that 100% of the broth tubes would develop turbidity. But this is not so, because if we substitute $m = 1.0$ into the Poisson equation (Eq. 6.2) we get, for $x = 0$:

$$P = \frac{e^{-1} . 1^0}{1!} = 0.3679, \text{ or approximately } 37\%$$

This means that if the bulk suspension contains 1.0 cfu per ml, randomly distributed, then 37% of 1.0 ml inocula will be free from bacteria. The other 63% of inocula will contain 1 or more infective particles which will give rise to turbidity. Thus, the ID_{63} dose (i.e. the dose that infects 63% of recipients) is the logical end-point for those infectivity titrations where one viable unit is sufficient to produce the observed response in the target.

In some very uniform systems the investigator may be lucky to find that the above simple assumptions correspond to reality. But in many other cases this will not be so. Thus, not all of the 'agents' will necessarily be equally effective and not all target sites equally sensitive. Therefore when investigating putative one-hit processes, it is important to establish which assumptions are acceptable and which are not. A good discussion of these points, with a variety of microbiological examples, is provided by Meynell and Meynell (1970, Chap. 6). Meanwhile Section 6.5 below describes how departures from the Poisson may be detected and dealt with.

6.4.2 Spontaneity of mutation to streptomycin resistance

This is not a Poisson-distribution problem but instead deals with the analysis of replicate bacterial colony counts for their homogeneity. So from a statistical point of view it is analagous to the example in Section 6.2.3 above where the χ^2-test was used to examine the homogeneity of bacterial cell counts from replicate fillings of the Helber chamber.

In 1948, Demerec in a classic paper showed that bacteria could acquire resistance to streptomycin by spontaneous mutation in the absence of a mutagen or of exposure to streptomycin itself. The method is simple and elegant and involves comparing the frequency of streptomycin-resistant (S_R) mutants in two types of culture: (A) replicate samples from a *single* culture grown up from a small inoculum to a final concentration of 1.3×10^8 bacteria per ml, with (B) individual samples from 20 *independently-grown* cultures from the same seed stock as used for (A) and likewise grown from small inocula to a final concentration of 1.3×10^8 bacteria per ml. Having grown up both types of culture, the numbers of S_R mutants were determined by plating equal volumes onto a selective medium containing 5 units streptomycin per ml which permitted growth of the S_R mutants but suppressed growth of the unmutated bacteria. Table 6.5 presents the colony counts and statistical analysis.

Table 6.5 The Demerec experiment designed to show that streptomycin-resistant mutants can arise spontaneously in a bacterial culture and without exposure to the antibiotic

Group A: Numbers of colonies[a] in 15 replicate samples from a *single* culture	Group B: Numbers of colonies[a] in the unreplicated samples from *20 independently grown* cultures
142, 155, 132, 123, 140, 146, 141, 137, 128, 121, 110, 125, 135, 121, 112.	67, 159, 135, 291, 75, 117, 73, 129, 86, 101, 56, 91, 123, 97, 48, 52, 54, 89, 111, 164.
N = 15	N = 20
E = 131.2	E = 105.9
χ^2 = 17.3	χ^2 = 550.3
d.f. = 14	d.f. = 19
P = $>10\%$	P = $\ll 0.1\%$

[a]After plating on nutrient medium containing 5 units streptomycin per ml.
From Demerec, M. (1948), *J. Bacteriol.*, **56**, 63–74. *Reproduced by permission of the American Society for Microbiology.*

In Group A we have 15 replicate counts from a single culture vessel, giving an average (E) count of 131.2 with which the actual counts (O) may be compared for homogeneity using the χ^2 formula (Eq. 6.1). This gives a found value of $\chi^2 = 17.3$ which, with 14 degrees of freedom (d.f.), corresponds to $P > 10\%$, i.e. there is no evidence of heterogeneity.

In contrast, in Group B, we have cultures produced and tested under the same conditions, except that they were grown in 20 separate vessels but all to the same final bacterial concentration of 1.3×10^8 cells per ml. However, the number of S_R mutants in the samples from each vessel showed considerable variation — from 48 up to 291, with an average of 105.9. When analysed by the χ^2-test, this degree of variation shows highly significant heterogeneity, reflecting the rare and random appearance of S_R mutants in the individual culture vessels during the period of incubation and bacterial growth. This experiment thus established that mutation to S_R was a random event that occurred independently of exposure to the antibiotic.

6.5 Where Poisson Fails to Rule

While there are certain well-established instances of count data to which the Poisson distribution applies (Helber slide, Coulter Counter, radioactivity), there are other types of count data to which it definitely does not. Therefore with an unknown system it is worthwhile to examine the nature of the distribution (as we did with the Helber slide counts in Section 6.2.4) to see if the assumption of the Poisson pattern is justified. If the assumption turns out to be *wrong*, this information may be valuable not only for avoiding errors in calculating SEM or confidence limits based on the Poisson formula, but also as an insight into the objects or events under study and how they interact with each other.

Departure from the Poisson distribution can arise from a variety of sources:

(1) lack of independence, i.e. lack of randomness in the distribution of the particles, objects or events in space or time;

(2) lack of uniformity in the units of area, volume or time which act as targets or receptacles for the particles, objects or events. An important input to this category are the volumetric errors in sample dispensing, e.g. in pipetting supposedly equal volumes of bacterial suspensions on to culture plates or of radioactive samples into scintillation vials.

The next section deals in some detail with point (1) above.

6.5.1 Contagion, aggregation, clustering, shoaling and general patchiness

Although it is rash to make sweeping generalizations in biology, I am now going to make one: organisms, macro and micro, *when living in their natural habitats* are commonly *not* Poisson distributed. Remember that an essential feature of the Poisson distribution is that the objects or events should be completely independent of each other: they should not interact in any way; they should neither attract nor repel each other; the presence of one organism in a site should neither predispose nor prevent the occurrence of another organism there, or nearby. This does not generally happen with living things in their natural habitats, and the string of words in the title to this section describes what is usually seen when organisms in nature are observed without disturbance; indeed, the very existence of collective nouns implies the widespread tendency, especially of animals, to occur in gaggles, flocks, herds, prides, packs, tribes, troops, shoals, etc. The same tendency may be seen with microorganisms when they are observed under conditions where individual cells are visible, e.g. when pieces of glass or transparent plastic are immersed in the sea and colonized by marine bacteria, the pattern of settlement shows aggregation or clustering rather than a purely random distribution. Indeed, the term 'patchiness' is part of the basic vocabulary of the ecologist and implies non-randomness of distribution.

The word 'contagion' in the title may seem out of place and deserves explanation. The classic example is the distribution of infected trees in a plantation. Initially, when only a few trees are infected, their distribution may be quite random, reflecting the haphazard transmission of the infective agent from some external source. However, if the disease is contagious, then each initially-infected tree acts as a focus of infection from which the disease spreads to the immediate neighbours. After a while, the distribution of infected trees in the plantation therefore shows a patchy rather than a random distribution. Thus, from a statistical point of view, what is called a contagious distribution may present a similar pattern of clustering to, say, the shoaling of fish, yet will have been arrived at through a completely different biological mechanism. Bliss (1967) presents data illustrating a contagious distribution in an infected orange grove in Brazil.

Taylor (1961), in a short but important paper, reports on the distribution of 24 highly diversified species of organisms in their natural habitats: his list ranges from earthworms in grassland soil, to shellfish in a sandy beach, to ticks on sheep, to virus lesions on bean leaves, to haddock in the sea. It is noteworthy that of the 24 species whose distribution in a habitat was analysed, 22 showed markedly non-Poisson distributions.

As we shall see below, the numbers of colonies of heterotrophic bacteria that develop on nutrient plates when viable counts are done on samples of lakewater may, or may not, follow a Poisson distribution, depending on the conditions for growth.

In short, one should regard the Poisson distribution as a mathematical idea to which the objects under investigation may, or may not, approximate in some degree. What we now have to do is to consider how departures from the Poisson are detected and characterized.

6.5.2   Under-distribution and over-distribution

As emphasized above, the Poisson distribution requires that the particles, objects or events be *randomly* distributed. There are two main forms of departure from randomness known as under-distribution and over-distribution, the latter being much more common with organisms in their natural habitats.

Under distribution is seen, for example, in the arrangement of atoms in a crystal lattice, or of apple trees in regular spacing in an orchard as viewed from an aircraft. With free-living organisms, a state of under-distribution would imply some kind of territorialism leading to a relatively constant spacing of different individuals or groups of individuals in the environment.

Over-distribution, in contrast, is the opposite phenomenon of aggregation, patchiness, contagion, etc.

With living things in their natural habitats there may well be both over- and under-distribution existing simultaneously, one on a macro scale and the other on a micro scale. For example, with fish that occur in shoals, there may be over-distribution if the equivalent of the grid square of the Helber chamber is, say, 1 km² of the sea bed. However, in the micro habitat of the shoal itself and taking a unit sampling volume of say 1 m³ there may well be under-distribution, since each individual fish maintains a fairly constant distance from its neighbours.

6.5.3 Over- and under-distribution of total cell counts

Turning now to the more quantitative aspects of under- and over-distribution, let us take some total cell counts in a Helber chamber as a convenient starting point. To keep things simple and to provide continuity with the previously-worked example (Section 6.2.4) which showed no significant departure from the Poisson, let us now consider two sets of fictitious data: in both sets, the average count per square is the same as for the Poisson example, i.e. $m = 2.5125$, arising from 201 cells counted over 80 squares. However, as shown in Table 6.6, the distribution of the cells over the squares is different. Notice how in the under-distribution set there is a relative concentration of values around $x = 2$ or 3. This is what is meant by 'under-distribution', in that relative to a random distribution there is a high frequency of squares with occupancy numbers close to the mean, i.e. they have been 'under-distributed' from the central value.

Table 6.6 Four sets of Helber slide data to illustrate under, over, actual and pure Poisson distributions: all 4 sets relate to 201 cells counted over 80 squares, giving an average count per square of $m = 2.5125$

Occupancy number of square (x-value)	Number of squares containing x cells:			
	Under-distribution	Actual distribution[a]	Pure Poisson disstribution	Over-distribution
0	4	4	6.49	31
1	2	18	16.30	3
2	39	23	20.47	6
3	23	15	17.14	8
4	8	12	10.79	10
5	4	5	5.41	13
6	0	2	2.27	6
$\geqslant 7$	0	1	1.13	3

[a]As reported in Table 6.3.

In contrast, the over-distributed set shows a much less well-defined peak than the random distribution, with a higher than expected incidence of squares that are either empty or have relatively large occupancy numbers. A graphical plot of the over- and under-distributions, together with the Poisson, is provided in Figure 6.3. χ^2-tests (method of Section 6.2.5) on the distribution in columns 2 and 5 of Table 6.6 show highly significant ($P < 0.1\%$) departures from the Poisson.

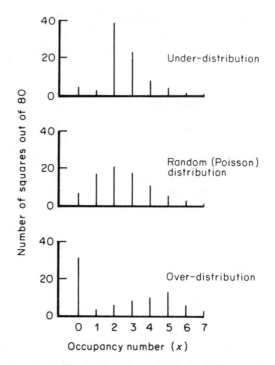

Fig. 6.3 Graphical plot of 3 distributions in each of which the average bacterial count per grid square of the Helber chamber is $m = 2.5125$. The centre plot is for a 'pure' Poisson distribution and is sandwiched between sets of data, taken from Table 6.6, that show under- and over-distribution

6.5.4 General procedures for dealing with departures from the Poisson

The main reason for wishing to determine the shape of the underlying distribution of count data is to allow application of the appropriate statistical procedures for summarizing and analysing the data. If, for example, as with radioactivity counts where there is a good assurance of Poisson data, the techniques appropriate to that distribution may be applied. If, however, the counts are non-Poisson, then the conscientious investigator has the choice of

(a) trying to find a suitable transformation of the data so that they can be summarized and analysed, or

(b) accepting that the nature of the underlying distribution is unknown and therefore applying non-parametric methods.

If we wish to pursue alternative (a), then the procedures described by Taylor (1961) should be followed. We have already stated that in a Poisson distribution the standard deviation (s) is equal to the square root of the mean count (m). More commonly this relationship is expressed in terms of the variance (s^2) as

$$s^2 = m \dots\dots\dots\dots\dots\dots\dots\dots\dots\dots\dots\dots\text{Eq. 6.3}$$

Taylor (1961) in his analysis of natural populations of 24 species of organisms showed that, in general, the divergences from the Poisson could be described by the equation:

$$s^2 = am^b \quad \dots\dots\dots\dots\dots\dots\dots\dots\dots\dots\dots \text{Eq. 6.4}$$

where a and b are constants that are characteristic of different species. For a Poisson distribution, $a = b = 1$, and the equation reduces to $s^2 = m$. In the populations studied by Taylor, a ranged from 0.035 to 3.0, and b ranged from 0.70 to 3.08.

The determination of a and b requires special assemblages of data, namely fairly large number of replicate observations at a series of different values of population density (m). Let us take heterotrophic bacterial viability counts as an example that will be of general interest to many microbiologists and ecologists.

Let us suppose that you are engaged on a routine programme of monitoring the heterotrophic bacterial count in samples of lakewater by a spread-plate method. Let us further suppose that equal portions of each water sample are distributed over, say, 10 replicate plates, so that the extent of plate-to-plate variation within a sample can be estimated with reasonable precision. This variation within each sample can be expressed as the variance (s^2) and tabulated opposite the value of mean count (m) for that sample. After the survey has been going for some time you should have an extensive collection of paired s^2- and m-values and will then be in a position to check the nature of the underlying distribution. The procedure is based on Eq. 6.4: $s^2 = am^b$. First, the equation is transformed to the logarithmic form:

$$\log_{10}(s^2) = \log_{10}a + b\log_{10} m \quad \dots\dots\dots\dots\dots\dots\text{Eq. 6.5}$$

Thus, a plot of $\log_{10}(s^2)$ against $\log_{10}m$ should give a straight line of slope b and intercept $\log_{10}a$. If the data follow a Poisson distribution, then $a = b = 1$. However, this may well not be the case. For example, Jones (1973) showed that total heterotrophic counts of bacteria in lakewater samples either followed or did not follow the Poisson distribution according to the exact conditions of culturing. Fig. 6.4, which is taken from Jones' article, illustrates this. Each graph is a double logarithmic plot of variance on the ordinate versus the mean count on the abscissa. Each point represents a single sample of lakewater which was plated out in several replicates so that an estimate of variance could be obtained. The best-fitting straight line was then imposed by the method of least squares (see Chapter 9). Thus, each graph provides values of a and b for Eq. 6.5.

If the viable counts were Poissonly distributed, $a = b = 1$ and each line should have a slope, b, equal to 1 and an intercept on the variance axis of $\log_{10}1 = 0$. In fact, however, none of the graphs shows this, although Fig. 6.4.c comes quite close.

What is the explanation for the departures from the Poisson which are displayed in Fig. 6.4? Why do the different conditions of incubation have the

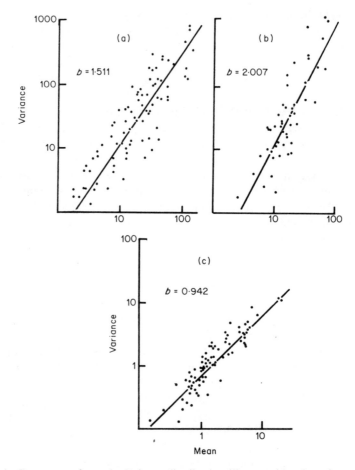

Fig. 6.4 Departures from the Poisson distribution illustrated by the colony counts of lakewater samples on nutrient agar plates. a, spread plate counts incubated for 24 days at 10°C; b, spread plate counts incubated for 15 days at 15°C; c, counts of microcolonies on membrane filters after incubation for 42 h at 15°C. *Reproduced by permission of J. G. Jones and the Society for Applied Bacteriology*

marked effects illustrated? Among the possibilities are (a) cross-feeding and (b) growth-inhibiting effects of neighbouring colonies on culture plates. When a bacterial cell lands on a nutrient agar surface it commences to grow and thereby alters the chemical composition of the culture medium in its vicinity. In a source such as lakewater there will be a mixed population of bacteria with different growth rates and nutrient requirements. It is perhaps not surprising, therefore, to find that the colonies, which may have arisen from single viable cells, do not behave as independent particles but exert influences on each other. We should not therefore be amazed to find departures from the Poisson distribution. Also, such departures are likely to be most pronounced if the plates are subjected to long incubation because it is under such conditions that diffusion of growth promoters and inhibitors will be most pronounced, and slow-growing organisms given the maximum opportunity of responding to such substances. It is noteworthy therefore that in Jones' data in Fig. 6.4, the closest approach to a Poisson distribution is provided by set c, where microcolonies on membrane filters were counted after less than 2 days incubation. Here, one would expect growth-interaction effects between bacteria to be least. In contrast, sets a and b with long incubation times of 15 and 24 days would favour the metabolic interactions along the lines discussed.

6.5.5 Transformation of the count variable

This section applies to both Poisson and non-Poisson count data and anticipates the material of the next chapter on Analysis of Variance. This technique, as we shall see, is only applicable to groups of data that are:

(a) normally, or approximately normally distributed, and

(b) have similar group-variances (but may have widely different group-means).

The second requirement means that count data cannot be subjected directly to analysis of variance because the variance is not constant, but increases as the count increases: to overcome this problem count data have to be transformed in an appropriate way so as to achieve normalization, and variances that are independent of the mean count.

For these purposes the commonly-used transformations are:

(1) a square-root transformation if the counts follow a Poisson distribution ($b = 1$);

(2) a logarithmic transformation, if $b = 2$;

(3) a negative fractional power transformation when $b = 3$.

The rule for making the transformation is that the raw data (y-values) should be raised to a power p which is defined by:

$$p = 1 - b/2 \dots\dots\dots\dots\dots\dots\dots\dots\dots\dots \text{Eq. 6.6}$$

where b is the slope of the line in the plot of log s^2 against log m. Thus, in Jones' data (Fig. 6.4) we can make the following transformations.

Set (a).

$$b = 1.511$$
$$\therefore p = 1 - 1.511/2 = 0.2445$$

which in practice, could be taken as $p = 0.24$. Thus, each original count (y-value) would be raised to the power 0.24 before any further reduction or analysis of the data were attempted. Raising to the power 0.24 can readily be done on a calculator with a y^x button.

Set (b).

$$b = 2.007$$
$$\therefore p = 1 - 2.007/2$$
$$p \simeq 0$$

This is to be interpreted as indicating the need for a logarithmic transformation of the count data.

Set (c).

$$b = 0.942$$
$$p \simeq 0.5 \text{ or } \tfrac{1}{2}$$

Here a square root transformation would be appropriate (i.e. raising each count to the power $\tfrac{1}{2}$)

Having made the appropriate transformation of the count data for the purpose of *normalizing* it, one can then apply the statistical methods appropriate to the normal distribution, i.e.

(1) using the arithmetic mean as a measure of central tendency;

(2) using the standard deviation as a measure of scatter;

(3) calculating the 95% confidence limits to provide an interval estimate of the underlying true population count;

(4) using the t-test to determine the significance of differences between the means of two groups;

(5) applying analysis of variance to analyse the sources of variation in more complex assemblages of data.

With (1), (2) and (3) the answers can be translated back into the scale of the original data by making the appropriate reverse transformation, e.g. taking antilogs if a log transformation was used.

7

Design of Experiments and Introduction to Analysis of Variance

An experiment is a device to make Nature speak intelligibly.

George Wald (1967) Nobel lecture.

If your experiment needs statistics, you ought to have done a better experiment.

Lord Rutherford (1871–1937)

(See Section 7.4.2 below for A.C.W. comment.)

7.1 More Mileage from a Simple Pipetting Experiment

7.1.1 Introduction to analysis of variance

In Chapter 1 we took a set of weight measurements of the amounts of water delivered by 3 1-cm^3 pipettes. Three replicate deliveries were taken from each pipette in a random sequence, and some introductory statistical exercises were performed on the 9 measurements obtained. This simple experiment is intended to exemplify a wide variety of laboratory procedures where replicate measurements are collected in groups and then need to be summarized and analysed. As the title of this section indicates, there is more statistical mileage to be obtained from these data. In particular, we can take them as material for analysis of variance (sometimes shortened to ANOVA or ANOVAR, but herein A. of V.) This procedure is one of the most powerful analytical tools in the whole of the statistician's repertoire, so let us apply it to the 9 pipetting results of Chapter 1 which are re-tabulated here for convenience. The 'working milligram' weights of deliveries from the 3 pipettes were:

Pipette A	Pipette B	Pipette C
37.5	11.6	24.8
16.8	24.7	27.7
22.7	12.0	5.9
Mean ($\bar{y}_A$) 25.67	Mean ($\bar{y}_B$) 16.10	Mean($\bar{y}_C$) 19.47

The question to be answered by A. of V. is basic: Are there significant differences between the mean weights of water delivered from the 3 pipettes? That is to say, do the mean values 25.67, 16.10 and 19.47 reflect *genuine volumetric differences* between the 3 bits of glassware, or is this amount of fluctuation only to be expected from the amount of variation in the replicate deliveries from each pipette?

7.1.2 Procedure for analysis of variance

Analysis of variance is done by a sequence of simple steps.

Step 1. Ensure that the data have been gathered in a randomized, or modified-randomized, sequence, or by some other method that avoids bias, i.e. the first step in A. of V. takes place *before the experiment is done:* the statistical procedures have to be built in *before* the data are gathered.

Step 2. Unscramble the randomization and arrange the data as columns of measurements under each group heading, i.e. pipette, in this case—as already done above.

Step 3. Add up the measurements in each column to obtain the group totals (*T*-values) as shown below:

	Pipette A		Pipette B		Pipette C
	37.5		11.6		24.8
	16.8		24.7		27.7
	22.7		12.0		5.9
Group total: $T_A =$	77.0	$T_B =$	48.3	T_C	58.4

Overall total $(\Sigma y) = 77.0 + 48.3 + 58.4 = 183.7$.

This introduces an important feature of the A. of V. technique, namely that although we are primarily concerned with differences between the group *means* ($\bar{y}$-values),what we actually work with in the analysis is the group *totals* (T-values).

Step 4. Identify the potential sources of variation in the data. This is the fundamental conceptual element in A. of V. In the present example we can identify *two independent* potential sources of variation which act together to produce the *total* variation. These two independent sources are:

(1) possible volumetric differences between the 3 bits of glassware, and

(2) operator variability, i.e. the sum-total of *unidentifiable* factors — mainly the wobbly hand and imperfect eye of the investigator — which cause supposedly identical manipulations to give replicate measurements that are *not* identical.

In general, where we have several groups of measurements with several replicates in each, we can list the 3 sources of variation as:

Total variation. This is the overall amount of scatter or dispersion in the whole collection of the observations, and ignoring temporarily that they belong to different groups. This total variation can be expressed quantitatively as the standard deviation, or variance, of all the readings lumped together as we did in Chapter 1.

Between-group variation. This is the amount of scatter that is represented by the variation *between* group means or group totals.

Within-group variation. This is the amount of scatter of individual observations (replicates) *within* each group (averaged over all the groups).

The essence of A. of V., i.e. where the *analysis* comes in, is that it provides an arithmetical procedure to subdivide the *total* variation into its two separate components:

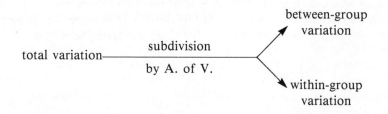

In order to do this subdivision of the total variation we have to subdivide first the *degrees of freedom* (d.f.) and then the *sums of squares* (s.s.).

144

Step 5. Subdivide the d.f. With $N = 9$ observations, we have a total of $N - 1 = 8$ d.f.

If we have k groups, the d.f. *associated with between-group variation* i.e. between pipettes, is $(k - 1)$. So here with 3 groups we have *2 d.f. associated with 'between-groups'.*

Degrees of freedom are additive and subtractible between the different sources; therefore if we know that the total d.f. = 8 and the between-group d.f. = 2, then the within-group d.f. must be $8 - 2 = 6$, by subtraction. We should have the following branch diagram in mind:

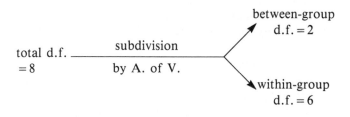

We can also arrive at d.f. = 6 for the within-group variation by recognizing that each group has 3 observations and therefore 2 d.f. 'inside' or 'within' it. Since there are 3 such groups, each containing 2 d.f., we have $3 \times 2 = 6$ d.f. *'associated with within-group variation'.*

We can generalize the assignment of the d.f. of data where there are k groups and a total of N observations, and assuming equal numbers of observations per group as:

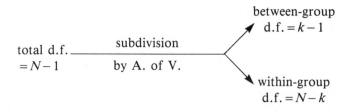

Note that $(k - 1) + (N - k) = N - 1$, which maintains the rule that the components of variation into which the total is split should have d.f.s that add up to the total d.f.

Step 6. We do a similar subdivision of the sums of squares (s.s.). Variance is defined as *sum of squares* divided by *degrees of freedom*, i.e. if we remove the square root from the formula for standard deviation we get:

$$\text{variance} = s^2 = \frac{\Sigma y^2 - (\Sigma y)^2/N}{(N - 1)} \dots\dots\dots\dots\dots\dots\dots\dots\dots \text{Eq. 7.1.}$$

The sum of squares (s.s.) constitutes the top line of this equation, i.e.

$$\text{s.s.} = \Sigma y^2 - (\Sigma y)^2/N$$

We have 3 s.s. to calculate: the total, the between-group and the within-group. They bear the same additive relationship to each other as do the d.f., i.e.

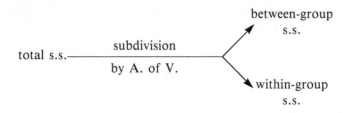

Thus, s.s., like d.f., are additive and subtractible. This means that if we calculate the total s.s. and the between-group s.s., we can get the within-group s.s. by subtraction. The total s.s. is given by:

$$\text{total s.s.} = \Sigma y^2 - (\Sigma y)^2/N \dots\dots\dots\dots\dots\dots\dots\text{Eq. 7.2}$$
$$= 37.5^2 + 16.8^2 + \dots + 27.7^2 + 5.9^2 - (183.7)^2/9$$
$$= 4509.57 - 3749.52$$
$$= 760.05$$

The between-group s.s. is given by:

$$\text{between-group s.s.} = \Sigma T^2/3 - (\Sigma y)^2/N \dots\dots\dots\dots\dots\dots\text{Eq. 7.3}$$
$$= \frac{77.0^2 + 48.3^2 + 58.4^2}{3} - (183.7)^2/9$$
$$= 3890.82 - 3749.52$$
$$= 141.30$$

The within-group s.s. is obtained not by applying a formula, but by subtracting the between-group from the total, i.e.

$$\text{within-group s.s.} = 760.05 - 141.30$$
$$= 618.75$$

An important point is that the divisor of the ΣT^2 term in Eq. 7.3 is determined by the *number of original observations that go into each* T-*value*, and *not* by the number of T-values. Thus if we had 5 groups with 2 observations in each, the divisor would be 2. This is dealt with again in Section 7.1.3 where generalized formulae for simple A. of V. are given.

Step 7. Construct an A. of V. table (Table 7.1). Take each column in turn and work systematically from left to right.

Columns 1, 2 and 3 have been explained in the text above. It is convenient to label the numerical value of the s.s. in column 3 with ①, ② and ③ as shown.

Table 7.1   Analysis of variance table for the 9 pipetting results

Source of variation	Degrees of freedom	Sums of squares	Variance (or mean square)	Variance ratio (F)
Total	8	$\Sigma y^2 - (\Sigma y)^2/N$ ..... ① $= 760.05$	—	
Between groups	2	$\Sigma T^2/3 - (\Sigma y)^2 N$ .... ② $= 141.30$	②/2 = ④ (70.65)	④/⑤ $= 0.68$[a]
Within groups	6	① − ② .... ③ $= 618.75$	③/6 = ⑤ (103.125)	

[a]Note that for insertion into the F-table, this particular ratio must be inverted so as to meet the requirement that the greater mean square (variance) must be in the numerator. We therefore calculate ⑤/④ = 1.46 with 6 and 2 d.f., respectively. Since the 5% tabulated value of F for 6 and 2 d.f. is 19.33, the found value of 1.46 is not significant.

Column 4, the variance, is the s.s. divided by the corresponding d.f., i.e.

$$\text{total variance} = ①/8;$$

but since we do not make any use of it, we do not bother to calculate it and the space is left blank as shown in the A. of V. table. Note that variance is sometimes referred to as *mean square*:

$$
\begin{aligned}
\text{between-group variance} &= ②/2 = ④ \\
&= 141.30/2 \\
&= 70.65 \\
\text{within-group variance} &= ③/6 = ⑤ \\
&= 618.75/6 \\
&= 103.125
\end{aligned}
$$

Column 5 contains the 'punch line' of the whole procedure. It contains the ratio of the between-group variance divided by the within-group variance and is given the symbol F, the variance ratio, which we have already met in Chapter 3.

$$F = \frac{\text{between-group variance}}{\text{within-group variance}}$$

$$= ④/⑤ = 70.65/103.125 = 0.68$$

If, as here, the F-ratio yielded by the data is less than 1.0, then we must take the additional step of calculating its reciprocal, since the F-tables in the Appendix will not accept F-values less than 1.0. Effectively therefore we have to calculate.

$$F = (5)/(4) = 1.46 \text{ with 6 and 2 d.f.}$$

instead of

$$F = (4)/(5) = 0.68 \text{ with 2 and 6 d.f.}$$

Step 8. Set up a null hypothesis which states that: 'the extent of between-group variation, as shown by the scatter of the *T*-values, is due to random-sampling fluctuations and does not reflect real differences in the 3 groups of measurements.'

Note that as with other null hypothesis (NH), and irrespective of the naked-eye appearance of the data, one asserts that *all* the variation shown by the data is due to random effects and that none is due to any of the groupings imposed by the investigator.

The NH is tested by comparing the found value of F, i.e. $F = 1.46$ with the tabulated values of F for 6 and 2 d.f. (Appendices A3a and A3b). These tabulated values are:

$$F = 19.33 \text{ for } P = 5\%$$

and

$$F = 99.33 \text{ for } P = 1\%$$

Since our found value of $F = 1.46$ falls far short of the $F = 19.33$ tabulated for significance at the $P = 5\%$ level, the NH is not rejected and we may conclude that 'the differences in pipette totals (or means) are easily explainable in terms of the amount of variation shown by the replicate readings obtained with each pipette.' We therefore have no grounds for supposing that the data indicate significant differences between the 3 bits of glassware. This does *not* mean that the 3 pipettes are *identical*, but only that any differences between them are too small to be demonstrable in the face of so much 'wobble' introduced by the operator using them.

7.1.3 Fiddling around with the data to get altered values of F

One of the best ways to get the 'feel' of A. of V. is to make some deliberate alterations to the raw data and see what effect it has on the final *F*-ratio and the NH.

Fiddle no. 1. Make the groups totals (*T*-values) *more diversified* by adding 10 to each pipette-A value and subtracting 10 from each pipette-B value. This yields the following set of 'fiddled' data:

	Pipette A	Pipette B	Pipette C
	47.5	1.6	24.8
	26.8	14.7	27.7
	32.7	2.0	5.9
T-value	107.0	18.3	58.4

Note that the overall total $\Sigma y = 183.7$ is the same as in the original data because of the symmetrical way in which the fiddling has been done.

148

Table 7.2 Analysis of variance of the 9 pipetting results subjected to 'Fiddle no. 1'

Source of variation	Degrees of freedom	Sums of squares	Variance	Variance ratio
Total	8	1934.05		
Between groups	2	1315.29	657.645	6.38[a]
Within groups	6	618.75	103.125	

[a]This result, being greater than 1.0, requires that the F-table be entered at vertical column 2 and horizontal row 6 to give tabulated F-values:

$$F = 5.14 \text{ for } P = 5\%$$
$$F = 10.92 \text{ for } P = 1\%$$

which is the opposite of what was done with Table 7.1.

The A. of V. gives the values recorded in Table 7.2. You will see that the within-group s.s. is exactly the same as for the unfiddled data (Table 7.1). This is what we should expect since the fiddling procedure did not alter the amount of scatter of the replicates associated with each pipette. More importantly, however, is the new F-value of 6.38 which corresponds to $P < 5\%$ but $> 1\%$, i.e. we have a 'significant' result. Remember that the P-value is the probability of that value of F arising if the NH is correct. If the P is sufficiently small (by convention $\leqslant 5\%$) we reject the NH and declare that 'random sampling fluctuations cannot reasonably be accepted as the explanation of the differences in pipette means and totals'. Thus, here there *is* evidence of genuine between-glassware variation that emerges as a clear 'signal' above the 'noise' of the operator variability.

Fiddle no. 2. Keep the group totals (T-values) the same as in the original data but reduce the amount of scatter within each column by adding 10.0 to the smallest value in each column and subtracting 10.0 from the largest value in each column. This gives:

	Pipette A	Pipette B	Pipette C
	27.5	21.6	24.8
	26.8	14.7	17.7
	22.7	12.0	15.9
T-values	77.0	48.3	58.4

Note that the T-values are exactly the same as in the unfiddled data.

Analysis of variance yields the values in Table 7.3. Observe that the between-group s.s. and variance are the same as in the unfiddled data, this being expected from the fact that the fiddling did not affect the group totals. However, the adjustments made within each group has the effect of greatly reducing the within-group variance from 103.125 in the unfiddled data to 17.79 in fiddle no. 2.

We then get a found value of $F = 3.97$, which does not quite reach the 5% point, although it comes much closer to it than does the value $F = 1.46$ yielded by the unfiddled data. In short, although in this 'fiddle no. 2' we have

Table 7.3 Analysis of variance of the 9 pipetting results subjected to 'Fiddle no. 2'

Source of variation	Degrees of freedom	Sum of squares	Variance	Variance ratio
Total	8	248.05		
Between groups	2	141.30	70.65	3.97
Within groups	6	106.75	17.79	

substantially reduced the 'noise' level of operator variability, it is still not quite low enough to make the 'signal' of possible between-glassware differences emerge at the conventional level of significance.

The analogy of the signal-to-noise ratio is helpful in understanding the output from the A. of V. The 'signal' is the deliberately-imposed subdivisions of the data implicit in the design of the experiment, while the 'noise' is the sum-total of unaccountable random variation mainly introduced (in this example) by the hand and eye of the operator.

An additional point of terminology: the *within-group* variation, which is due in this experiment mainly to the wobbly hand and imperfect eye of the human operator, has its counterpart in all A. of V. tables. It is often referred to as *residual* variation or *error* variation. The word *error* does not imply any mistake, but simply the sum-total of unaccountable random effects that cause replicate observations to vary.

7.2 General Aspects of Simple Analysis of Variance

7.2.1 The procedure summarized

By 'simple' analysis of variance (A. of V.), is meant that which can be applied to N observations (y-values) arranged in k equal-sized groups and with N/k observations in each group. This type of A. of V. is sometimes referred to as 'one-way analysis of variance' or 'single factor ANOVA'. As emphasized in Section 7.1.1, it is essential that the observations be gathered in a randomized or other unbiased way. The summarized procedure for A. of V. is:

Step 1. Tabulate the N observations in an array with k vertical columns which have N/k y-values in each. Add up the y-values in each column to get the group totals (T-values).

Step 2. Using the formulae of Table 7.4, carry out the A. of V, working with each column in turn and proceeding from left to right, finishing up with a single found value of F. Note that the denominator of the F-ratio is always the within-group variance, except occasionally when the reciprocal has to be taken if the found F-value is below 1.0.

Step 3. Assess the significance of the found F by comparing it with the tabulated values of F for $P = 5\%$ and 1%, entering the F-table at the degrees of freedom (d.f.) indicated in the footnote to Table 7.4.

Table 7.4   General formulae for simple analysis of variance of N observations arranged as k equal-sized groups

Source of variation	Degrees of freedom	Sum of squares	Variance	Variance ratio
Total	$N-1$	$\Sigma y^2 - (\Sigma y)^2/N \ldots$ ①	—	—
Between groups	$k-1$	$\dfrac{k\Sigma T^2}{N} - \dfrac{(\Sigma y)^2}{N}$ .. ②	②$/(k-1)$ $=$ ④	④ / ⑤
Within groups	$N-k$	① $-$ ② $=$ ③	③$/N-k$ $=$ ⑤	—

Note: To interpret the variance ratio (found value of F) in column 5, compare it with the tabulated values of F for $P=5\%$ and 1%, entering the F-table with $(k-1)$ d.f. across the top horizontal row and $(N-k)$ in the left-hand vertical column. Exceptionally, if ④/⑤ yields a value less than 1.0, then the ratio ⑤/④ must be calculated and the F-table entered with $(N-k)$ d.f. across the top horizontal row and $(k-1)$ d.f. in the left-hand vertical column.

Step 4.   If the found F is less than the tabulated F at the $P=5\%$ level, there is no significant differences between the Groups.

If the found F is greater than or equal to the tabulated F at the $P=5\%$ or $P=1\%$ levels, the differences between groups are declared respectively to be 'significant' or 'highly significant'.

7.2.2   Pre-conditions

Like all statistical procedures, A. of V. has certain conditions which must be fulfilled to ensure validity. These are:

(1) Unbiased gathering of the data. As emphasized above, it is essential that randomization, or some equivalent assurance against biasing, be incorporated into the actual conduct of the experiment and the collecting of the data.

(2) The underlying population(s) from which the groups of data are drawn must be normally distributed, or approximately so.

According to statisticians, the A. of V. technique is 'robust' towards minor departures from normality, and many investigators seem to take this as an excuse for not bothering to find out whether the data are normally distributed or not.

Normality may be checked by the rankit or probit procedures outlined in previous chapters (if you have deliberately collected enough data of the right kind). If the data are known to be not-normally distributed, then either a suitable normalizing transformation must be introduced (Section 8.4.1) or non-parametric methods used (Section 8.4.2).

(3) The variances within each group should not be significantly different from each other. This may be checked by Bartlett's Test outlined below.

(4) Replication must be built into the experiment design, i.e. A. of V. is not possible with only one observation per group.

Although not essential, it is convenient for simplicity of formulae if there are equal numbers of observations in each group. Sometimes, however, symmetry is spoiled accidentally because an observation is lost during the experiment. If this happens, then there are several possible courses of action.

(1) To discard at random one observation from all the other groups in the experiment and simply accept the wastage of information, with consequent reduction of sensitivity of the F-test through loss of d.f. However, this is less satisfactory than

(2) To adjust the formulae in the A. of V. to take account of the lost observation. For example, suppose in the data we considered above under 'fiddle no. 1', the third observation of pipette B was lost so that we were left with:

Pipette A	Pipette B	Pipette C
47.5	1.6	24.8
26.8	14.7	27.7
32.7	lost	5.9
$T_A = 107.0$	$T_B = 16.3$	$T_C = 58.4$

Table 7.5 Analysis of variance of pipette data to allow for a lost observation (see text)

Source of variation	Degrees of freedom	Sum of squares	Variance
Total	7	$\Sigma y^2 - (\Sigma y)^2/N$ .... ① $= 1552.71$	—
Between pipettes	2	$\dfrac{T_A^2}{3} + \dfrac{T_B^2}{2} + \dfrac{T_C^2}{3} - \dfrac{(\Sigma y)^2}{N}$ .... ② $= 959.17$	479.58
Within pipettes	5	① $-$ ② $= 593.54$ .... ③	118.71

Variance ratio $F = 479.58/118.71 = 4.04$.
Tabulated F for 2, 5 d.f. $= 5.79$.
Therefore, the apparent between-pipette variation is not significant.

The A. of V. would be as set out in Table 7.5. Note that there are changes in the d.f. and in the s.s. for between-pipettes to allow for two of the T^2-values having a divisor of 3 and one to have a divisor of 2. The final output of the A. of V. is $F = 4.04$ which with 2, 5 d.f. is not significant at the $P = 5\%$ level. This illustrates how with a small experiment of initially only 9 observations the loss of a single observation may be quite important and may lead to a different conclusion from that reached with a full set of data (the 9 observations of 'fiddle no. 1' gave $F = 6.38$, which with 2, 6 d.f. was significant at the $P = 5\%$ level).

(3) See Section 8.3.5 for formulae to replace a mising value.

7.2.3 Internal consistency of the F- and t-tests

If the experiment contains only 2 groups of measurements, and if we are interested in finding out whether the difference in the two means is significant, then we can do either a t-test or an A. of V. and F-test. Let us now check that the two procedures do in fact give the same result from quite different formulae and are thus internally consistent. Let us take as our working example the following 6 observations.

	Pipette A	Pipette D
	37.5	59.1
	16.8	39.9
	22.7	47.4
$T =$	77.0	146.4
$\bar{y} =$	25.67	48.8
$s^2 =$	113.723	93.63
$N =$	3	3

First, apply the t-test (Eq. 3.1):

$$t = \frac{(\bar{y}_D - \bar{y}_A)\sqrt{N}}{\sqrt{(s_A^2 + s_D^2)}}$$

$$= \frac{(48.8 - 25.67)\sqrt{3}}{\sqrt{(113.723 + 93.63)}}$$

$$= 2.78$$

Consulting the t-table for $2N - 2 = 4$ d.f. we find that the 5% point of $t = 2.78$. Thus, the difference between the means of the 2 groups of measurements is judged by the t-test to be *just* significant at the $P = 5\%$ level. (To let the reader into a secret, the data were deliberately chosen to give this result exactly on the tabulated 5% point of t.)

Table 7.6   Analysis of variance of the measurements from pipettes A and D

Source of variation	Degrees of freedom	Sum of squares	Variance	F
Total	5	$\Sigma y^2 - (\Sigma y)^2/N \ldots$ ① $= 1217.43$		
Between pipettes	1	$\dfrac{\Sigma T^2}{3} - \dfrac{(\Sigma y)^2}{N} \ldots$ ② $= 802.73$	802.73 $=$ ④	④ / ⑤ $= 7.74$
Within pipettes	4	① $-$ ② $= 414.70$ ③	103.67 $=$ ⑤	

Now, let us do an A. of V. and F-test: from the calculations in Table 7.6 we get a found $F = 7.74$. Consulting the F-table at d.f. $= 1, 4$, we read out the tabulated value from $P = 5\%$ as 7.71, i.e. the same as the found F, within rounding-off errors. Thus, the A. of V. procedure delivers the same conclusion as the t-test, namely, that the *means* of the two groups of data are just significantly different at exactly the $P = 5\%$ level. Therefore the t-test and the A. of V. procedure are alternative and mutually consistent methods for assessing the significance of differences in the *means* of two groups of data. Perhaps we should not be surprised at this since the t-test and the A. of V. procedure both require that the data:

(1) be normally distributed,

(2) be obtained by random sampling,

(3) show no significant differences in the standard deviations of the replicates in each of the two groups.

However, whereas the t-test is restricted to comparing the means of only 2 groups, the A. of V. procedure allows the comparison of as many groups as you like. It also permits the analysis of grouped data with multiple criteria of classification, as we shall see below; so it is much more versatile than the t-test.

A final note: the homogeneity of standard deviations (or variances) of the groups of measurements subjected to t-testing is assessed by making an F-test (see Section 3.6.1), whereas the corresponding procedure to accompany A. of V. is Bartlett's test which is dealt with next.

7.2.4 Bartlett's test for homogeneity of variances

This is superficially complicated but is actually not difficult. Let us go back to our 9 'original' pipetting results (Section 7.1.1) and, taking each column in turn, calculate the standard deviation, s. However, we do not actually want the three s-values themselves, so while we have each of them on the dial of the calculator, calculate s^2 and then $\log_{10} s^2$. This gives us

	Pipette A	Pipette B	Pipette C
	37.5	11.6	24.8
	16.8	24.7	27.7
	22.7	12.0	5.9
s^2	113.7233	55.5100	140.1433
$\log_{10} s^2$	2.05585	1.74437	2.14657

Note that the values of s^2 and $\log s^2$ are recorded to 4 or 5 decimal places.

Step 2. Add up the s^2 values horizontally, i.e.

$$\Sigma s^2 = 113.7233 + 55.5100 + 140.1433$$
$$= 309.3766$$

then divide this by the number of groups (pipettes) $= k = 3$ to give the *mean within-pipette variance*:

$$\Sigma s^2/k = 103.12553$$

and take the $\log_{10}$ of this value to get:

$$\log_{10}(\Sigma s^2/k) = 2.01337$$

Multiply this by the number of groups (pipettes) $= k = 3$ to give:

$$\begin{aligned} k\log_{10}(\Sigma s^2/k) &= 3 \times 2.01337 \\ &= 6.04011 \end{aligned}$$

For simplicity, designate this as A for use in Bartlett's formula, below.

Step 3. Refer back to the table of pipette readings and the bottom line where the $\log_{10}s^2$-values are recorded. Add these up to get:

$$\begin{aligned} \Sigma \log_{10}s^2 &= 2.05585 + 1.74437 + 2.14657 \\ &= 5.94679 \end{aligned}$$

For simplicity, designate this as B for use in Bartlett's formula.

Step 4. Apply Bartlett's formula:

$$M = 2.3026\, f\,(A - B) \dots\dots\dots\dots\dots\dots\dots\text{Eq. 7.4}$$

where M is Bartlett's statistic and f is the number of degrees of freedom associated with (i.e. 'within') each group (pipette in this example). Here with 3 observations per group, $f = 2$. Substituting our previously found values for A and B we get:

$$\begin{aligned} M &= 2.3026(2)(6.04011 - 5.94679) \\ &= 0.42976 \end{aligned}$$

Step 5. Treat M as a χ^2 statistics with $k - 1$ degrees of freedom. The null hypothesis states that 'each value of s^2 is an estimate of the same underlying σ^2, i.e. that the variances are homogeneous.' If, therefore, the value of M is less than the tabulated value of χ^2 for $k - 1 = 2$ d.f. at the $P = 5\%$ level, then the NH is not rejected. This is the case here where the tabulated χ^2 for 2 d.f. and $P = 5\%$ is 5.99.

If M approaches, or is greater than, the tabulated χ^2 for $k - 1$ d.f. and $P = 5\%$, then an additional refinement in the calculation of the test statistic is required (it is not needed in this example but we shall do it just the same).

Step 6. Following on from the above, calculate a correction factor (C):

$$C = 1 + \frac{k + 1}{3kf} \dots\dots\dots\dots\dots\dots\dots\text{Eq. 7.5}$$

where $k =$ the number of groups, here $k = 3$, and $f =$ the number of d.f. associated with each s^2-value. This gives

$$C = 1 + \frac{3+1}{3(3)(2)}$$

$$= 1.2222$$

Step 7. Calculate a corrected test statistic M/C. Here

$$M/C = 0.42976/1.2222$$
$$= 0.352$$

Enter the value of M/C into the χ^2 table and interpret the result as in Step 5. Since C is always greater than 1.0, the effect of using M/C instead of M is to lower the test statistic and thereby allow acceptance of some marginal hypotheses which otherwise might be rejected.

Summary. To apply Bartlett's test to k groups of data each consisting of $(f+1)$ observations, calculate:

$$M = 2.3026 \, f \, [k \, \log_{10}(\Sigma s^2/k) - \Sigma \, \log_{10}s^2] \, \ldots . \text{Eq. 7.6}$$

If M approaches, or exceeds, the tabulated value of χ^2 for $k-1$ d.f. and $P = 5\%$, calculate M/C as described in Steps 6 and 7 above. If the found value of M or M/C is less than the tabulated value of χ^2, the variances may be taken as acceptably homogeneous, or at least not significantly heterogeneous.

If the data consist of groups with unequal numbers of observations, the Bartlett test requires a somewhat more complicated formula (see Snedecor and Cochran, 1967, p. 296).

7.2.5 *Potential confusion in terminology*

The term 'analysis of variance' could be considered something of a misnomer in that the *purpose* of the analysis is to assess the homogeneity or heterogeneity of group *means*; furthermore the arithmetical *method* of the analysis is to split up sums of squares (s.s.) and degrees of freedom (d.f.) rather than to analyse variances directly. Variances only come in towards the end of the procedure when one calculates the variance ratio F. This author thinks that it might have been better to call the procedure 'analysis of variation of group means'. This would then leave the way clear to call Bartlett's test 'analysis of variances' (in the plural) — which is what *it* is concerned with. However, like most semantic discussions, this one is likely to be fruitless except to point out that the term 'analysis of variance' can be misleading, since the actual analytical separation is done on *d.f.* and *s.s.* and the purpose of the procedure is to compare the *mean values* of the different groups into which the data fall.

7.3 Regrettably We Now Need Some More Jargon

Some readers may feel that they have had their fill of the weights of water delivered by 1-cm^2 pipettes. However, no special apology is made for sticking

156

to the pipette example (a) because it has such a direct relevance to many types of quantitative data that experimental biologists collect frequently with some kind of measuring instrument, and (b) because for purposes of understanding diverse statistical techniques, there is merit in sticking to a single kind of easily visualized system. But perhaps, enough is enough, and we should now start to diversify.

Anonymous
> . . . like the statistician who was drowned in a lake of average depth six inches.

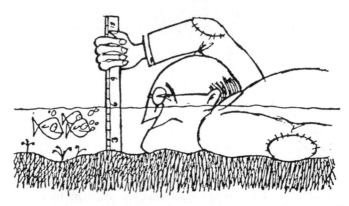

From MacKay, 1977, and reproduced through the courtesy of the Institute of Physics

7.3.1 The agricultural influence

Because certain important statistical methods evolved in response to the needs of agricultural scientists, many of the special jargon terms relating to experiment design have farming or horticultural connotations, so they may sound quaint to the laboratory experimentalist.

Let us start with the words *plot* and *field*. These come straight from agriculture where the field is literally an area of ground capable of supporting the growth of crops. In order to do an experiment in the field, the ground has to be divided up into *plots*, each of which is large enough to accommodate one, or several, plants whose growth or fruit yield, etc. is to be measured. Thus, the plot is the *basic unit of observation* and corresponds to the *individual weighing* of a delivery of water from a pipette.

Having got our field divided into plots, we are now ready to apply *treatments*. These could be, for example, different fertilizers applied to plots sown with the same seed. But equally, the 'treatment' could consist of different varieties of seed in the various plots, all given the same fertilizer. Or we could have a more complicated experiment still, where the field contains several varieties of, say, wheat, each treated with several different fertilizers. In this case we would have 2 *independent* variables, or *factors* as they are called, that might affect the yield from a plot. Each *combination* of these factors would then be '*a treatment*'. The investigation of the fertilizer + wheat variety effects could be made still more elaborate by not only using different mixtures of fertilizers but applying each of them at different *levels* (doses or frequencies). Thus, in what is called a *factorial* experiment, we may have several *independent variables*, or *factors*, such as fertilizer type, fertilizer level (dose or frequency) strain of wheat and perhaps time of planting. The experiment should be set up in such a way that each *combination* of the variables is represented as a different *treatment*, and we shall have to ensure that there are enough plots to cater for this and also to permit replication of each combination (treatment).

Because the natural fertility, drainage, shelter from wind, access to sun, etc. in different parts of our field may vary, we shall have to assign the various *treatments* to *plots* in such a way as to *avoid systematic error*. For example, it would be wrong to have all the replicate plots of a particular treatment clustered together in one part of the field. Instead, the replicates of the various treatments should be distributed in such a way that any variations in underlying soil fertility are not unwittingly made coincident with one of the deliberately applied treatments. We can achieve this in various ways.

(1) A fully randomized design, in which the various treatments are allocated to plots in an entirely random arrangement. This, however, has the potential drawback that, by chance, all the replicates of a particular treatment may find themselves clustered together. We might therefore be better to use:

(2) The randomized block design. Here we first subdivide the field into fairly large areas of equal size, which we call *blocks*, and then we subdivide each *block* into *plots* of such a size and number that we have one plot for each treatment. Within each block, the allocation of treatments to plots is done in a random fashion, hence the name randomized block. If the blocks are big enough, we may decide to have more than one representative of each treatment combination within each block; such replicates would then be described as 'nesting within the blocks'.

(3) Fields being what they are, we may know from past experience that the underlying fertility is non-uniform and that, for example, there is a gradient of fertility from the top end of the field, which is well-drained, to the bottom end which is boggy. Simultaneously, there may be a gradient of wind exposure operating at right angles to the drainage gradient. If, therefore, we are forced to do our experiment in such a location, then the randomized block design might not be quite as good as the Latin square, which is essentially a double system of blocking applied simultaneously.

(4) The Latin square. Suppose our experiment consists of 6 different treatments, A, B, C, D, E, and F, with 6 replications of each; then we divide up the field into a 6×6 grid pattern of, say, 6 east–west rows and 6 north–south rows. We then assign the treatments in such a way that each letter comes up once in each east–west row and once in each north–south row, e.g.

```
B A C F D E
E D F C A B
D B E A C F
F C A B E D
C F D E B A
A E B D F C
```

This arrangement does not *eliminate* the natural gradients of fertility or wind exposure, but it does help to ensure that any effects of such gradients are not wrongly attributed to an effect of one of the treatments (such as might happen in a purely random design). Fisher and Yates (1963) present Latin squares from 4×4 up to 12×12 in their Table XV. Squares of the appropriate size should be chosen at random.

Fields are never of infinite size and there is therefore likely to be a *natural constraint* on how many plots can be accommodated.

Thus, using the agricultural idiom, we have introduced the terms *plot, block, field, nesting, treatment, factor, level, row, natural gradient* and *constraint*. Let us now start to apply these jargon terms to experiments in the laboratory.

7.3.2 A weight is a plot, a day is a block and a pipette is a treatment

We saw in Chapter 2 that statisticians have taken certain ordinary English words and infused them with special meanings for statistical purposes, thereby creating jargon. Let us now apply some of the above agricultural statistical jargon to the pipetting experiment where each weight-measurement corresponds to a *plot*, each pipette to a *treatment* and the whole experiment to a *field*. Suppose that if instead of collecting the 9 readings at one sitting, we collected 3 readings — one from each pipette — on each of 3 successive days, then each day's work would constitute a *block*. If, on each day, the sequence of pipette usage was independently randomized, then the experiment would be a *randomized block design*. Suppose, furthermore, that on each of the above days we took *two* measurements from each pipette, then such pairs would be described as 'nesting within blocks' or a *split-plot* design. Now, perhaps, you can make some sense out of the jargon-laden title of this section 'a weight is a plot, a day is a block and a pipette is a treatment'.

7.4 Experiment Design as an Art-Form

7.4.1 What do we mean by experiment design?

Experiment design may briefly be described as the system of rules for allocating *treatments* to *experimental units*. Experiments *need* to be designed, in the sense that they do not just happen. There has to be a positive act of intellectual creation which emerges as 'a device to make Nature speak intelligibly' (George Wald, 1967 Nobel lecture). Thus, the raw materials of an experiment are the experimental *units* or *subjects*, together with the question or questions being asked by the investigator, plus the physical facilities needed to do the work. We can list quite simply the basic components of experiment design.

(1) To have a clearly formulated question, or series of questions.

(2) Randomization: to make sure that possible sources of bias are reduced or eliminated by adopting one of the schemes of randomization discussed above.

(3) Replication: there should be at least two experimental units assigned to each treatment so that the extent of random variation, residual variation or error can be quantified.

(4) Blocking: the experiment design should take account of any natural sub-divisions of the material; or, it may be expedient to divide up a large 'field' into blocks.

(5) Symmetry: it is highly desirable that the numbers of experimental units assigned to each treatment be uniform, otherwise the analysis of variance may become unnecessarily difficult or even impossible. In complex experiment designs, such as bioassays, additional elements of symmetry may be needed.

(6) Perfection versus feasibility: some compromise usually has to be struck between the design demands of statistical perfection and the realities of actually doing the experiment *on* the material, and with the physical resources, including manpower, that are available.

It can therefore be seen that experiment design is indeed an art-form in that elements of intuitive selection, allocation of emphasis, symmetry and economy all play their part. It is a considerable intellectual challenge to put together, as an integrated package, the various components of experiment design so that what emerges is an economical and logical framework that maximizes the output of useful information and minimizes the possibilities of false conclusions or worthless results. Well-designed experiments are also aesthetically satisfying.

7.4.2 Rutherford revisited

Good experiment design may well yield conclusions that are so obvious as to make laborious statistical analysis scarcely necessary. In this sense Rutherford was right when he said that 'if your experiment needs statistics, you ought to have done a better experiment' — except that he was using 'statistics' to mean only the calculations done after the data have been gathered. However, the message in this book is that the 'statistics' have to be built in *before* the data are collected, and thus a 'good' experiment is necessarily one which incorporates at least *some* statistical principles such as replication and randomization. Without these, the work may be little more than 'a pseudo-experience in para-science', and finish up being published in the *Journal of Irreproducible Results* or the *Acta Artefacta*!

7.4.3 Recognizing the constraints

Experiments, in practice, often have to be built around unavoidable physical constraints imposed by the available number of experimental units, or by

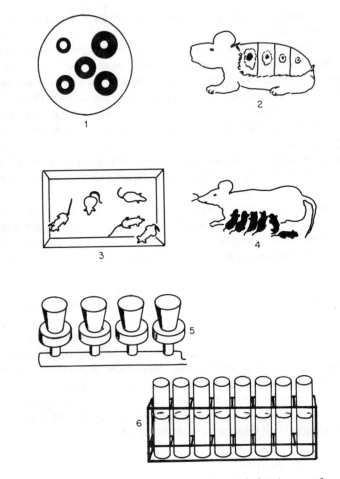

Fig. 7.1   Examples of physical constraints in experiment design (see text for comments)

equipment, personnel or time. Common constraints encountered in biological work are illustrated in Fig. 7.1

(1) Petri dishes, as used for antibiotic assays or immunodiffusion tests, will only accommodate at the most about 6 or 8 wells or disks from which test substances can diffuse and produce zones of growth-inhibition or precipitation. Where possible, therefore, the individual petri dish should constitute a 'block' and the experiment should contain at least two such blocks to allow random variation to be assessed. Each block should contain one of each of the 'treatments' being studied in the experiment.

(2) A shaved guinea-pig being used to test a dermonecrotic toxin or for dermal allergy studies. As with the petri dish, the individual animal should constitute a 'block' and will probably accommodate 8 or 10 test areas ('plots'). Since the skin may not be uniform in its responsiveness, e.g. show some kind of gradient from front to rear or from mid-line to flank (like the boggy field), the allocation

of treatments to 'plots' should use either randomized block or Latin square designs.

(3) Mouse experiments. Usually the individual cage, holding up to about 20 mice of the same age and sex, will constitute the 'block'. Each cage should contain one or more animals assigned to each treatment, and further replication (if necessary) assured by having two or more cages.

(4) A litter of newborn mice being used to test an infectious or toxic agent. Here the litter is the 'block', with each baby a 'plot'. Since there will be uncontrollable variation in the numbers of babies in different litters, the 'blocks' will be of varying size (if maximum use is to be made of the material). Also, the numbers of males and females will vary and this may have to be taken into consideration if sex might influence the results. Investigators wanting to use litters of newborn animals, and to use them with maximum efficiency, are probably going to employ *incomplete randomized block* designs and should consult a statistician for expert advice. These designs may also be useful in the agar-plate and guinea-pig skin systems mentioned above.

(5) Multiple-filtration apparatus for collecting microbial cells in metabolism studies with radioactive tracers. Here both the *in vitro* age of the cell suspension and the duration of exposure to the radio-labelled metabolite will be important variables (in addition to pH, temperature, co-factors, etc). Unless the investigator has an army of slaves, there is a limit (imposed by his own dexterity and by the number of filters on the apparatus) to how many samples can be run in close succession. Here the 'block' consists of a series of treatments which can be set up and processed with each loading of new filters into the filter-holders. If possible, each such loading should permit one 'plot' of each treament. Randomization of sequence and filter-holder should be incorporated to neutralize possible bias due to the bacteria ageing or to physical irregularities in individual filters. Replication should be introduced by running two or more blocks.

(6) Rack of test tubes being used to set up a colorimetric assay such as the Lowry test. In most such tests the colour takes time to develop, and in some tests it then fades. Since it is not possible to prepare and read a large number of tubes absolutely simultaneously, a time constraint may restrict the number of tubes that it is reasonable to set up in a particular assay. Bias might be introduced unless randomization is applied both in the sequence of preparation of the test mixtures and in the sequence in which they are read spectrophotometrically. Thus, the 'block' will consist of a row of tubes containing the various amounts of standard and unknowns arranged and processed in a random sequence; and the assay should contain at least two such blocks, run in tandem.

In contrast to most of the above, our 9-result pipetting experiment had very few constraints. There was no obvious physical or other logistical restriction on the exact numbers of pipettes which could be tested at one sitting, or in the number of replications obtainable with each. In such circumstances, a fully randomized design, as actually used, was satisfactory. If, however, there had been reason to suspect an intermittent fatigue, or training, effect influencing the performance of the investigator, then it might have been better to have used a randomized block design in which 3 measurements were taken, one from each pipette in random sequence, after which there was a pause. Then another 3 measurements would be taken and then a pause; and the final 3. The analysis of variance of a randomized block design is given in the next chapter.

8

Experiment Design
and Analysis Continued

When I was a child, I spake as a child, I understood as a child, I thought as a
child; but when I became a man, I put away childish things.

The First Epistle of Paul to the Corinthians (xiii, 2)

8.1 From Pipettes to Mice

This chapter gets away from pipettes and introduces a set of measurements of the concentrations of glucose in the sera of 24 mice which had been subjected to 4 different experimental treatments in a pharmacological investigation. Each treatment was administered to 6 mice and each mouse yielded a single specimen of serum on which a glucose determination was made. We first consider the experiment in its simplest form and then progressively introduce additional elements of design, such as blocking and nesting, to show how they affect the analysis of variance and the output of information. The experiment was actually done by the method to be described last which had all the sophistication built into it; however, for purposes of explanation it is best if we start simple and then progressively build in the various design features one by one. Thus, we shall be using the same set of 24 serum-glucose measurements throughout as convenient data to illustrate a variety of design elements. Let us start with the simplest presentation of the data (Table 8.1) which shows 4 columns, each with 6 results. Also included are the total, mean and standard deviation for the results in each column.

Table 8.1 Glucose concentrations (mg/100 ml) in the sera of mice that had been subjected to 4 different experimental treatments A, B, C and D

	A	B	C	D
	221	94	330	163
	200	109	302	157
	233	146	283	177
	180	141	273	139
	198	124	307	148
	213	114	279	144
Total (T)	1245	728	1774	928
Mean ($\bar{y}$)	207.5	121.33	295.67	154.67
S.D. (s)	18.7910	19.7754	21.4445	13.9809

8.1.1 One-way analysis of variance

The results in Table 8.1 follow the pattern already discussed in detail in the previous chapter (Sections 7.1 and 7.2). They exemplify the 'fully-randomized design with one experimental variable'. The analysis of variance will therefore follow the procedure for 'one-way A. of V. with equal-sized sets' or 'single-factor ANOVAR'. It should be noted that the blood was collected in a properly randomized sequence as was, prior to that, the allocation of animals to treatments. It should also be mentioned that blood-glucose levels in mice are approximately normally distributed (Furman et $al.$, 1981). Table 8.1 shows that the standard deviations for the 4 groups are not vastly different; they do in fact pass the Bartlett test for homogeneity of variances (Section 7.2.4). Therefore the criteria for analysis of variance (A. of V.) are fulfilled, in that we have random samples of a normally-distributed variable and with homogeneity of the group variances.

The next step is to work out the subdivision of the degrees of freedom:

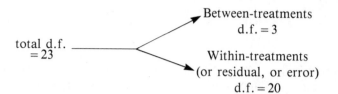

total d.f.
= 23

Between-treatments
d.f. = 3

Within-treatments
(or residual, or error)
d.f. = 20

Table 8.2   Analysis of variance of the data of Table 8.1

Source of variation	Degrees of freedom	Sum of squares	V	F
Total	23	$\Sigma y^2 - (\Sigma y)^2/N$ ..... ① = 111,057.9583	—	—
Treatments	3	$\Sigma T^2/6 - (\Sigma y)^2/N$ ... ② = 104,060.4583	34,686.82	99.14**
Residual	20	① − ② = 6997.5 .. ③	349.875	—

$s = 18.7049$
Tabulated F for d.f. 3,20 = 3.10 ($P = 5\%$) and 4.94 ($P = 1\%$).
Blanks are left where the result is of no interest.

The A. of V. is then set out in Table 8.2 which follows from the explanations given in the previous chapter. We have, however, started to introduce some abbreviations, i.e. in column 1 we have the word 'treatments' which is to be understood as meaning 'between treatments'; this saves a lot of repetitious writing of the word 'between'. And also in column 1 we just use the simple word 'residual' for 'within treatment' or 'error' variation: remember that 'error' has a special meaning in this context. It does not infer any mistake, but simply the sum-total of unassignable effects that cause individual mice to yield different values of serum glucose even though the animals were selected for uniformity at the start of the experiment and then handled in as consistent a manner as possible in the administration of each treatment.

The square root of the residual variance (column 4) is the within-group, or error, standard deviation and is given as a footnote of the table ($s = 18.7049$). This value of s can also be calculated from the 4 individual group standard deviations in Table 8.1. Note that to obtain a value for average standard deviation of the measurements within 4 groups such as A, B, C and D, it is incorrect simply to take the arithmetic mean of the 4 s-values. If you have k equal-sized groups, with individual s-values s_1, s_2, etc., then the 'average' standard deviation (s) is given by:

$$s = \sqrt{\frac{(s_1^2 + s_2^2 + \ldots s_k^2)}{k}} \quad \ldots\ldots\ldots\ldots\ldots\ldots\ldots\ldots\ldots\ldots Eq. 8.1$$

166

i.e.
$$s = \sqrt{\frac{1}{4}(18.7910^2 + 19.7754^2 + 21.4445^2 + 13.9809^2)}$$

$$= \sqrt{349.875}$$

$= 18.7049$ (the same as obtained as a by-product of the A. of V. table).

Returning to Table 8.2, column 5 contains the ratio of between-treatment variance ÷ error variance = 99.14. The double asterisk is the conventional shorthand way of indicating that this F-value is higher than that tabulated for $P = 1\%$ and the appropriate number of d.f. (here: 3, 20). A single asterisk indicates significance at the $P \leqslant 5\%$ level and 'NS' signifies a non-significant result, i.e. $P > 5\%$ and with the null hypothesis remaining unrejected.

8.1.2 Supplementary t-testing

There are certain types of experiment where one of the treatments is a control, or baseline, procedure and where the main interest is in identifying which of the other treatments have yielded results that are significantly different from this control value. The one-way analysis of variance, as just done, informs us that there is *heterogeneity* in the treatment means, but it does not tell us *which* particular groups are significantly different from which others. Suppose, therefore, that we identify treatment A as the control treatment with which the others are to be compared, one at a time. The problem then boils down to doing a series of t-tests in which we compare the means of treatments B, C and D with that of treatment A. However, instead of actually making these multiple comparisons, we can define instead the boundaries around the control mean outside of which any other mean of 6 observations would be significantly different at the $P = 5\%$ or 1% levels.

The t-test formula for comparing the means of two groups with equal numbers of observations has already been given as:

$$t = \frac{(\bar{y}_2 - \bar{y}_1)\sqrt{N}}{\sqrt{(s_1^2 + s_2^2)}} \quad \dots\dots\dots\dots\dots\dots\dots\dots\dots\dots\dots\dots\dots\dots\text{Eq. 3.1}$$

where N is the number of observations in each group ($= 6$ in this example).

Let us identify the two $\bar{y}$-values as: $\bar{y}_1 = $ mean of the control group (in this example 207.5); and $\bar{y}_2 = $ the mean that another group has to have to be just significantly different at the $P = 5\%$ or 1% levels.

Instead of s_1^2 and s_2^2, which represent the variances of the two groups being compared, we can use s^2 from the bottom line of the A. of V. table (8.2) since this is the overall within-group variance for the whole experiment. So $(s_1^2 + s_2^2)$ can be replaced by $2s^2 = 2 \times 349.875$. We can then rearrange Eq. 3.1 as:

$$\bar{y}_2 = \bar{y}_1 \pm \frac{t}{\sqrt{N}} \cdot \sqrt{2s^2} \quad \dots\dots\dots\dots\dots\dots\text{Eq. 8.2}$$

where the $\pm$ is introduced to indicate that we want boundaries of significance both above and below the control value.

We need the tabulated values of t for $2N - 2 = 10$ d.f. These are 2.228 for $P = 5\%$ and 3.169 for $P = 1\%$. Substituting the first of these into Eq. 8.2 gives:

$$\bar{y}_2 = 207.5 \pm \frac{2.228}{\sqrt{6}} \sqrt{2 \times 349.875}$$

$$= 207.5 \pm 24.0608$$
$$= 183.4 \text{ and } 231.6$$

To get the boundaries corresponding to $P = 1\%$, we substitute $t = 3.169$ and get:

$$\bar{y}_2 = 207.5 \pm 34.223$$
$$= 173.3 \text{ and } 241.7$$

If we now refer back to the treatment means for Groups B, C and D, we see that they all lie outside of the boundaries defined even by the $P = 1\%$ criterion. Thus, they are all highly significantly different from the control group that received treatment A.

A number of precautions are required when using the t-test in this way. In the first place, statisticians tend to frown on multiple comparisons, on the grounds that if you have enough groups, then through random-sampling fluctuations alone, you will likely get one that falsely appears to be significantly different from the control. This objection can be partly overcome by:

(1) Identifying in advance, before the data have been gathered, which group is the control, i.e. do not just arbitrarily pick the lowest or the highest mean after the data have been gathered.

(2) Taking a larger tabulated value of t than one normally would for a t-test with 2 groups, e.g. take the $P = 1\%$ value of t for a difference to be labelled as 'significant' and $P = 0.1\%$ for a difference to be labelled as 'highly significant'.

A further point relates to the actual setting up of the experiment and the use of the control group as the focal point for multiple comparisons as set out in Section 3.5.1.

8.2 Two Independent Variables

8.2.1 The experiment revealed

Staying with the same data, let us now consider what was actually done to the mice.

Groups B and D. These groups were given on day 0 (when the animals were 3–4 weeks old) an infecting dose of 2×10^5 colony-forming units of *Bordetella pertussis*, the whooping cough bacillus, so as to establish a sub-lethal pulmonary infection in the animals.

Groups A and C. On the same day 0, these groups were instilled with bacterial diluent fluid (casamino acid solution) and therefore provided a set of no-infection controls. Fourteen days later, all the mice were given a subcutaneous injection of either adrenaline (500 µg/kg body weight) in acidified saline (Groups C and D) or were given acidified saline alone (Groups A and B). A blood sample was then taken 10 minutes later from each animal under ether anaesthesia and the serum separated and analysed for glucose.

The experiment was thus concerned with examining the effect, on serum glucose, of *two independent variables*, i.e. pertussis infection (as compared with no infection) and adrenaline (as compared with saline).

The first step in the new analysis of variance is therefore to put more informative headings at the top of each column of results in place of the code letters, A, B, C and D. We shall not re-tabulate the 24 observations, but simply give the *T*-values under the appropriate headings:

Saline on day 14		Adrenaline on day 14	
Uninfected A	Pertussis B	Uninfected C	Pertussis D
$T_A = 1,245$	$T_B = 728$	$T_C = 1,774$	$T_D = 928$

8.2.2 Constructing an interaction table

The interaction table is a convenient way of setting out the *T*-values (treatment totals) for purposes of some extra totalling operations. It is worthwhile always to construct an interaction table whenever you are dealing with two, or more, independent variables. Here it is with the *T*-values in place and the row totals and column totals inserted:

	Uninfected	Pertussis	Total
Saline	$T_A = 1,245$	$T_B = 728$	$T_A + T_B = D_1$ $= 1,973$
Adrenaline	$T_C = 1,774$	$T_D = 928$	$T_C + T_D = D_2$ $= 2,702$
Total	$T_A + T_C = B_1$ $= 3,019$	$T_B + T_D = B_2$ $= 1,656$	$\Sigma y = 4,675$

Just as we introduced the symbol *T* for the treatment totals, so we have introduced *B* (for the bacterial variable) for the *column* totals for uninfected and pertussis-infected mice, and *D* (for the drug variable) for the *row* totals for mice given saline or adrenaline.

8.2.3 Allocating degrees of freedom

Here we have a branching which starts out the same as before, except that additionally we can now subdivide the 3 degrees of freedom (d.f.) associated with *treatments*. This subdivision is into 3 categories (sources of variation).

Pertussis. In respect of this variable we have 2 'levels', i.e. no-infection $(A + C)$ and infection $(B + D)$ with totals B_1 and B_2. Since there are two 'levels', there is 1 d.f. We can also express it by saying that there is '1 d.f. associated with pertussis'.

Adrenaline. Likewise in respect of *this* variable, we have 2 levels, i.e. no-drug $(A + B)$ and drug $(C + D)$ with totals D_1 and D_2, and therefore '1 d.f. associated with drug'.

This leaves us with 1 d.f. apparently surplus. It is referred to as 'the interaction term'. This term occurs whenever we have 2 or more *independent* variables. We can label it alternatively as 'pertussis × adrenaline', or 'the interaction between drug and infection'. We shall see exactly what it signifies as we progress with the analysis. Meanwhile the allocation of d.f. (omitting the word 'between') is:

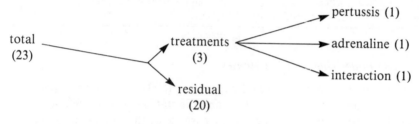

8.2.4 Analysis of variance and F-tests

The analysis of variance (A. of V.) table with the 'treatments' subdivided in 3 ways is just an extension of Table 8.2. Since the raw data and the T-values are unaltered, the only new things needed are those parts of the analysis relating to pertussis, adrenaline and interaction.

Reference to Table 8.3 shows that the sums of squares (s.s.) for pertussis and adrenaline are formed in the usual way by squaring the appropriate totals, adding up the squares and dividing by the number of original observations (12) that contribute to each B- or D-value.

The interaction s.s. is not calculated by formula, but is had by subtracting the pertussis and adrenaline s.s. from the treatment s.s. This follows from the additivity and subtractability properties of s.s.

The final output of the analysis is the 3 F-ratios in column 5, all of which are double-asterisked, indicating that they are highly significant. As is normally the case, the denominator of each F-ratio is the residual or error variance. Note that we are no longer interested in the F-ratio for the treatments term now that we have resolved it into its 3 sub-components. The 3 F-ratios relate to 3 underlying null hypotheses as follows.

Table 8.3   Analysis of variance of the serum glucose data, but with the two independent variables of pertussis and adrenaline included (formulae previously given in Table 8.2 are not repeated)

Source of variation	Degrees of freedom	Sum of squares	V	F
Total	23	$111,057.9583 \ldots \ldots ①$	—	—
Treatments	3	$104,060.4583 \ldots \ldots ②$	—	—
Pertussis	1	$\dfrac{\Sigma B^2}{12} - \dfrac{(\Sigma y)^2}{N} \ldots ③$ $= 77,407.042$	77,407.042	221**
Adrenaline	1	$\dfrac{\Sigma D^2}{12} - \dfrac{(\Sigma y)^2}{N} \ldots ④$ $= 22,143.375$	22,143.375	63.3**
Interaction (pertussis × adrenaline)	1	$② - ③ - ④ =$ $4510.04 \ldots \ldots ⑤$	4510.04	12.9**
Residual	20	$① - ② = 6997.5 \ldots ⑥$	349.875	—

Tabulated F for d.f. $1,20 = 4.35$ (for $P = 5\%$) and 8.10 (for $P = 1\%$).
Blanks are left where the result is of no interest.

8.2.5   The underlying null hypotheses

The setting up of one or more null hypotheses (NH) is an intrinsic part of a test of significance, such as the F-test, and the statement that one makes is in no way determined by the naked-eye appearance of the raw data, i.e. even if it is patently obvious that groups X and Y are different, one still goes through the motions of asserting that the apparent difference is attributable purely to random sampling fluctuations.

Our first NH is that relating to pertussis and its effect on serum-glucose levels. We have 12 observations from uninfected mice (total $B_1 = 3,019$, mean = 251.58) and 12 from infected mice (total $B_2 = 1,656$, mean = 138.00). The NH states that 'the difference in the mean serum glucose of 251.58 (mg/100 ml) in uninfected mice as compared with 138.00 in pertussis mice is due solely to random sampling fluctuations in the 2 groups, each of 12 observations', i.e. the NH postulates that there is no 'real' difference in serum-glucose levels between normal and pertussis-infected mice. By 'real' difference we mean that there are two underlying populations of serum glucose concentrations with different mean values (μ_1 and μ_2).

The test of the NH is the F-ratio, which with a value of 221 is vastly greater than the tabulated value of F for d.f. $= 1, 20$ ($F = 8.10$ for $P = 1\%$). Therefore we can say that '*if* the NH is correct, the probability of getting the results we got is vanishingly small'. Therefore we reject the NH and assert that 'There is a highly significant difference in mean levels of serum glucose in control and pertussis-infected mice'.

Turning to the effect of adrenaline, we have a similar NH and similar firm rejection of it, in that the average glucose level of the 12 adrenaline-treated mice was 225.17, while that of saline-injected mice (lumping together the infected and uninfected as was done for adrenaline) was 164.42. Again, this difference turns out to be highly significant.

We are now left with the third NH to deal with, namely that relating to 'interaction'. The easiest way to state it is: 'There is no interaction of variable 1 on variable 2 other than that due to random-sampling fluctuations'. But this does not really explain what interaction is. If we just set the explanation aside for a moment, the A. of V. in Table 8.3 shows that there is highly significant interaction ($P<1\%$). So what is this thing called 'interaction'?

Let us take uninfected mice given saline (Group A) as 'normal, baseline' animals: they had a mean glucose value of 207.5.

Look at the uninfected group given adrenaline (Group C): they had mean glucose of 295.67, i.e. adrenaline caused an increase of 88.17 in the level of serum glucose. Now switch attention to infected mice. Such animals, when given saline (Group B), had serum glucose of 121.33. Let us now suppose there was *no* significant interaction. We would then expect that the giving of adrenaline would increase the glucose in these infected mice by the same absolute amount of 88.17 ($\pm$ sampling fluctuations) as it did in the uninfected mice. So we would *predict* that Group D (pertussis + adrenaline) should (in the absence of interaction) have a glucose level of 121.33 + 88.17 = 209.5. Instead of which the actual level of Group D was 154.67. So what we mean by 'significant interaction' is a matter of non-additivity of the effects of the 2 independent variables.

Looking at it another way, the giving of adrenaline to uninfected mice caused the blood glucose to go up by 88.17, whereas the giving of adrenaline to pertussis-infected mice only caused it to go up by 154.65−121.33 = 33.32. The A. of V.

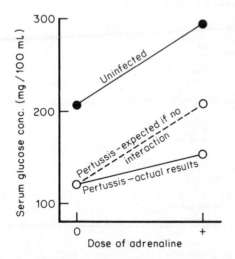

Fig. 8.1   Pictorial representation of the results of the pertussis/adrenaline experiment.
Each point represents the mean response of a group of 6 mice

172

and F-test allow us to assess objectively whether the 2 increments, 88.17 and 33.32, could reasonably have arisen by sampling fluctuations, or whether pertussis-infected mice are inherently less responsive to adrenaline (in respect of its hyperglycaemic affect) than uninfected mice. The NH asserts that 'the difference between the 2 increments is due solely to sampling fluctuations'. But the high value of F causes this NH to be rejected with firmness.

We can therefore summarize the findings as follows:

(1) Pertussis-infected mice were highly significantly *hypo*glycaemic relative to controls.

(2) Adrenaline caused highly significant *hyper*glycaemia in both pertussis-infected and uninfected mice.

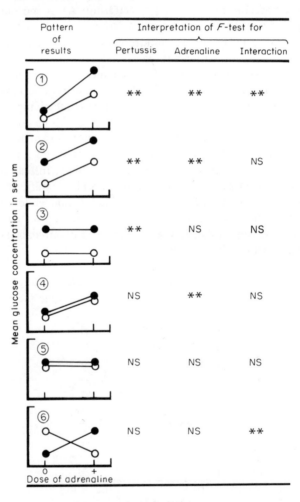

Fig. 8.2 Pictorial representation of the main possible patterns of pertussis/adrenaline results compared with the corresponding outputs from analysis of variance ($\bullet$ = uninfected; $\circ$ = infected; ** = highly significant; NS = not significant)

(3) Pertussis-infected mice showed a highly significant *attenuation of adrenaline-mediated hyperglycaemia* compared with uninfected animals.

The value of A. of V., when applied routinely to experimental data, is that it provides standard criteria for assessing the effect of experimental variables when these effects may tend to be partially obscured by the inescapable variability of biological systems. It may also expose interactions of various kinds, which without A. of V. might not be apparent.

8.2.6 Pictorial equivalents of analysis of variance

It is useful to visualize the output from the A. of V. as corresponding patterns of raw data. For example, Fig. 8.1 is a diagrammatic representation of the pertussis/adrenaline results. The 4 treatment-group means are represented as 4 points on graphs of serum-glucose concentration plotted against dose of adrenaline (0 or +). The two types of pre-treatment (pertussis or no-infection) are represented as solid or open circles, and we get a diagram of 2 non-parallel lines. This pattern of results corresponds to an A. of V. where the 3 components—pertussis, adrenaline and interaction—are all highly significant (**).

Now inspect in Fig. 8.2 a series of analogous diagrams with different patterns of the two lines and with corresponding differences in the output from the A. of V. (These diagrams represent *possible* patterns of results, not results as actually obtained with pertussis and adrenaline.)

Diagram 1 is the actual results as above, and with all 3 terms of the A. of V. highly significant.

Diagram 2 shows the two lines parallel; correspondingly, the interaction term is now *not* significant (NS) but the other two are.

Diagram 3 shows an experiment where the adrenaline had no effect, but there was still a difference between pertussis-infected and uninfected animals. Therefore the *F*-test for pertussis gave a ** result but adrenaline and interaction did not.

Diagram 4 shows two sloping superimposed lines. This means that pertussis was without influence (NS) on glucose levels but adrenaline did produce a highly significant change (**). The parallelism of the lines means that the interaction is NS.

Diagram 5 shows all 4 groups with the same glucose level. Therefore neither pertussis nor adrenaline produced any changes. Likewise there is no interaction.

Diagram 6 shows the pattern corresponding to a ** interaction term but with both pertussis and adrenaline NS.

In A. of V. of real-life data one is going to encounter a variety of patterns intermediate between those shown here. Nevertheless in interpreting the output of an A. of V. on two independent variables, it is useful to have in mind these main patterns of *possible* results.

In conclusion, the reader should regard pertussis and adrenaline as just two examples of independent variables. Also, in place of serum-glucose concentrations in mice you should exchange the measurement variable of most interest in your own experimental work.

8.3 Superimposing Randomized Blocks with Nesting

8.3.1 Adding the blocks and nests

The experiment as outlined above assumed a fully-randomized design for its execution and analysis. This is not, in fact, exactly what was done. In particular, the withdrawal of blood took about 1 minute per mouse, which was done exactly 10 minutes ± 10 seconds after injecting the saline or adrenaline. Therefore it was convenient to take the mice in groups of 8, i.e. 'blocks' of 8, each block containing 2 mice from each of the 4 treatment groups. Such duplication within blocks is referred to as 'nesting' (Section 7.3.1), and so we had 4 'nests' (of 2 observations each) within each block. The animals within each block were randomized and injected with either saline or adrenaline, according to a predetermined scheme, with a 1-minute spacing between mice. This allowed each animal to be bled exactly 10 minutes after being given the injection and then there was a 2 minute break for the operators. Each block was therefore completed in injection and bleeding before the next block was started.

Table 8.4   Results of the pertussis/adrenaline experiment as actually done in randomized blocks with internal nesting. The various totals and subtotals are also given

Block no.	Saline on day 14				Adrenaline on day 14				Block total (K)
	Uninfected (A)		Pertussis (B)		Uninfected (C)		Pertussis (D)		
I	221 200	421[a]	94 109	203	330 302	632	163 157	320	1576
II	233 180	413	146 141	287	283 273	556	177 139	316	1572
III	198 213	411	124 114	238	307 279	586	148 144	292	1527
Treatment total (T)	1245		728		1774		928		4675

[a]This and the other nest-totals are given the symbol J.

Table 8.4 sets out the results (the same data that we have already been considering) but showing the actual arrangement of blocks and nests after unscrambling the randomization. The table also shows nest totals (J) and block totals (K). The treatment totals (T) are the same as before, as also is the interaction table (Section 8.2.2) showing how the T-values may be combined in different ways to give 'pertussis' totals (B-values) and 'adrenaline' totals (D-values). Having done all these totallings, sub-totallings and total-totallings, we are ready to put some extra branches into the degrees-of-freedom (d.f.) diagram.

8.3.2 A more elaborate degrees-of-freedom diagram

The rule in constructing a d.f. diagram is to first identify, in the hierarchy of groupings, the type of group which contains the smallest number of observations. Previously a 'treatment', being a group of 6 observations, was the smallest hierarchical category. But now the 'nest' is the smallest group. Therefore our first branch in the d.f. diagram is where we split the total variation into between-nests (which we write simply as 'nests', with 11 d.f. since there are 12 nests) and within-nests (which we label as 'residual'). We then get:

We then recognize that 'nests' are classifiable in 2 independent ways into blocks (with 2 d.f. since there are 3 blocks) and treatments (with 3 d.f. since there are 4 treatments). We also have 2×3 d.f. = 6 associated with the interaction of blocks and treatments. This gives us:

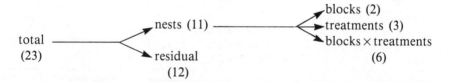

We then have a subdivision of treatments into pertussis, adrenaline and pertussis × adrenaline, so that the complete d.f. diagram is:

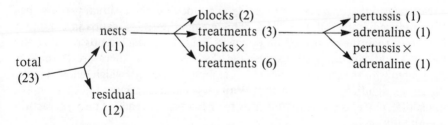

This is the structure of the experiment as actually done by Furman *et al.* (1981).

8.3.3 Analysis of variance and F-tests

The A. of V. is as set out in Table 8.5 which uses much of what has already been calculated in Tables 8.2 and 8.3. The only new components are those associated with the nesting and blocking. Note how the sum of squares (s.s.) of the blocks × treatments interaction term is obtained by subtraction and by following the logic of the d.f. diagram.

Table 8.5 Analysis of variance of the pertussis/adrenaline experiment, with blocking and nesting included (formulae previously given in Tables 8.2 and 8.3 are not repeated)

Source of variation	Degrees of freedom	Sum of squares	V	F
Total	23	111,057.9583 ...... ①	—	—
Nests	11	$\Sigma J^2/2 - (\Sigma y)^2/N$ ... ② $= 107,563.4583$	—	—
Blocks	2	$\Sigma K^2/8 - (\Sigma y)^2/N$ ... ③ $= 185.0873$	92.5	0.32[a] (NS)
Treatments	3	104,060.4583 ...... ④	—	—
Blocks × treatments	6	② − ③ − ④ $= 3317.913$ .... ⑤	552.98	1.90 (NS)
Pertussis	1	77,407.042 .... ⑥	77,407.0	266**
Adrenaline	1	22,143.375 .... ⑦	22,143.4	76**
Pertussis × adrenaline	1	④ − ⑥ − ⑦ $= 4510.04$ .... ⑧	4510.0	15.5**
Residual	12	① − ② $= 3494.5$ .. ⑨	291.2083	—

$s = 17.0648$
Tabulated F for d.f. 2, 12 = 3.89 for $P = 5\%$; 6.93 for $P = 1\%$.
Tabulated F for d.f. 6, 12 = 3.00 for $P = 5\%$; 4.82 for $P = 1\%$.
Tabulated F for d.f. 1, 12 = 4.75 for $P = 5\%$; 9.33 for $P = 1\%$.

[a]Note that for significance testing, this ratio must be inverted, i.e. 291.208/92.5 = 3.15 with 12 and 2 d.f., respectively. This is still not significant since the tabulated F for 12, 2 d.f. are 19.41 and 99.42 for $P = 5\%$ and 1%, respectively.

Notice also that as a result of nesting and blocking, the residual variance has been reduced from 349.9 to 291.2. This means that the F-ratios in column 5 are larger than the corresponding ones in Table 8.3, which tends to make the test of significance more sensitive. On the other hand, there are fewer d.f. associated with the error variance (12 as compared with 20) which tends to make the test of significance *less* sensitive. With these particular data, it does not make much difference, because the F-ratios for pertussis, adrenaline and interaction are already so large.

What about the value of the blocking as a useful or necessary part of the experiment design? The A. of V. shows that there were no significant differences between the blocks. This is a desirable result because it means that the 3 batches in which the mice were processed did not differ much from each other. However, if there *had* been significant block variation, such as might arise if the adrenaline solution deteriorated during the time of the experiment, it would have been very worthwhile knowing about it. Also, it would have been worthwhile to subtract it from the residual variation, of which it might otherwise have been an undesirably large part.

8.3.4 Blocks without nests

In recent sections we have been steadily making the pertussis/adrenaline experiment more complex by revealing additional features of the experiment as it was actually done. Let us now back-track to the slightly simpler design of randomized blocks without nests. This means that within each block there is only one representative of each treatment instead of the two or more representatives associated with nesting.

Let us stick to the same 24 serum-glucose values, but now imagine that the mice were processed in 6 randomized blocks, each containing one representative of each treatment. The results would then be as in Table 8.6, which shows the 6 block-totals (Q).

Table 8.6   Results of the pertussis/adrenaline experiment tabulated as if the data had been gathered as 6 randomized blocks without nesting

| Block no. | Saline on day 14 | | Adrenaline on day 14 | | Block total (Q) |
	Uninfected (A)	Pertussis (B)	Uninfected (C)	Pertussis (D)	
I	221	94	330	163	808
II	200	109	302	157	768
III	233	146	283	177	839
IV	180	141	273	139	733
V	198	124	307	148	777
VI	213	114	279	144	750
Treatment total	1,245	728	1,774	928	4,675

The block totals here are given the symbol Q to distinguish them from the larger blocks (symbol K) used in Tables 8.4 and 8.5.

The d.f. diagram starts by recognizing that we have 24 bits of data that are classified in 6 horizontal rows and 4 vertical columns, corresponding to the 2 independent variables, blocks and treatments. We therefore have the start of the d.f. diagram as:

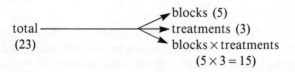

$$\begin{array}{l} \text{blocks (5)} \\ \text{total} \longrightarrow \text{treatments (3)} \\ \text{(23)} \qquad\qquad \text{blocks} \times \text{treatments} \\ \qquad\qquad\quad (5 \times 3 = 15) \end{array}$$

We do not have a 'residual' term in the same way as in the design with nesting. Instead, the blocks × treatments interaction term with 15 d.f. serves this function and we can label it as such.

We can then go on to subdivide the treatments as previously, so that the whole d.f. diagram is:

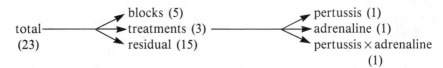

The corresponding A. of V. is set out in Table 8.7, which shows no significant block-variation, but has the pertussis, adrenaline and interaction components appearing highly significant, as before. Note that the residual variance at the bottom of column 4 is slightly different from that in the other A. of V. tables. This simply reflects the different assumptions inherent in the various experiment designs and is yet another example of the 'central message' that the exact method of analysis of data depend on how the data were gathered. And a randomized block design without nesting is not the same as *with* nesting, and neither of them is the same as a fully randomized design, even though in all 3 designs we have the same 24 results.

Table 8.7   Analysis of variance of the results in Table 8.6 (formulae previously used in Tables 8.2 and 8.3 are omitted)

Source of variation	Degrees of freedom	Sum of squares	V	F
Total	23	111,057.9583 ...... ①	—	—
Blocks	5	$\Sigma Q^2/4 - (\Sigma y)^2/N$ ... ② = 1880.7083	376.14	1.10 (NS)
Treatments	3	104,060.4583 ...... ③	—	—
Pertussis	1	77,407.042 .... ④	77,407.04	227**
Adrenaline	1	22,143.375 .... ⑤	22,143.37	64.9**
Pertussis × adrenaline	1	③ – ④ – ⑤ = 4510.04 .... ⑥	4510.04	13.2**
Residual	15	① – ② – ③ = 5116.7917	341.12	

Tabulated F for d.f. 5, 15 = 2.90 ($P = 5\%$) and 4.56 ($P = 1\%$).
Tabulated F for d.f. 1, 15 = 4.54 ($P = 5\%$) and 8.68 ($P = 1\%$).

8.3.5 Dealing with missing values

One of the differences between the fully-randomized design and the randomized block is that the former can accept groups containing different numbers of results, while the latter cannot, i.e. A. of V. of randomized-block data requires equality of group numbers. Sometimes, however, a single observation may accidentally be lost, and a procedure has therefore been developed to cope with this vexatious contingency. It consists of calculating a value (y_{ji}') for the missing observation by the formula:

$$y_{ji}' = \frac{tT_i' + bB_j' - G'}{(t-1)(b-1)} \quad \ldots\ldots\ldots\ldots\ldots\ldots \text{Eq. 8.3}$$

This formula also provides an opportunity for introducing the subscript notation which is standard format in textbook accounts of A. of V. but which so far has been avoided here.

Table 8.8   General table of data for a randomized block design, showing use of double-subscript notation

Block no.	1	2	Treatment no. 3	...	i	...	t	Block total
1	y_{11}	y_{12}	y_{13}	...	y_{1i}	...	y_{1t}	B_1
2	y_{21}	y_{22}	y_{23}	...	y_{2i}		y_{2t}	B_2
3	y_{31}	y_{32}	y_{33}	...	y_{3i}	...	y_{3t}	B_3
$\vdots$	$\vdots$	$\vdots$	$\vdots$		$\vdots$		$\vdots$	$\vdots$
j	y_{j1}	y_{j2}	y_{j3}	...	y_{ji}	...	y_{jt}	B_j
$\vdots$	$\vdots$	$\vdots$	$\vdots$		$\vdots$		$\vdots$	$\vdots$
b	y_{b1}	y_{b2}	y_{b3}	...	y_{bi}	...	y_{bt}	B_b
Treatment total	T_1	T_2	T_3	...	T_i ...		T_t	$\Sigma y = G$

Note that the subscript, e.g. for the top left y-value, is to be read as 'y-one-one' and not as 'y-eleven'. The 'one-one' means the intersection of the 1st horizontal row and the 1st vertical column. Similarly, y_{ji} is the result that is located in the jth row and the ith column.

In order to use the formula you have to build the mental picture of a general randomized block table (Table 8.8) which can have any number of treatment *columns* and any number of block *rows*. The treatment totals (T-values) are labelled with subscripts to give T_1, T_2, T_3, etc. up to T_t, the last column, while similarly the block totals are labelled B_1, B_2, B_3, etc. up to B_b, the last block. Any particular column between 1 and t is indicated by the symbol i and the total for this column is T_i; i can have any value between 1 and t. Likewise, any selected block between 1 and b is given the subscript j, so that we have B_j for that block total. Any particular y-value can be identified by giving it the double subscript ji, so that y_{ji} means the y-value in the jth block and the ith treatment. The symbol y_{ji}' (with the 'prime') means the *missing* y-value in the jth block and the ith treatment. T_i' means the total of the values that are left in the ith treatment column from which y_{ji}' is missing. Similarly, B_j' means the total of the values that are left in the jth block from which y_{ji} is missing. G' is the grand total (Σy) of all the observations actually gathered. With each of these symbols, the 'prime' superscript indicates that there is a missing value to be taken note of.

We are now ready to use the formula. Let us apply it to the results in Table 8.6 and let us suppose that the value 307 was lost from Treatment C of Block V. We therefore have:

t = number of treatments = 4,
T_i' = the T-value of what is left in the column with the lost result = 1467,
b = number of blocks = 6,
B_j' = the B-value of what is left in the row with the lost result = 470,
G' = the grand total of the results actually tabulated = 4368.

Substituting into Eq. 8.3 gives:

$$y_{ji}' = \frac{4 \times 1467 + 6 \times 470 - 4368}{(4-1)(6-1)}$$

$$= 288$$

This, then, would be the value we would insert as a replacement for the 'missing' one. The A. of V. is then carried out as though y_{ji}' were a real observation, except that the total sum of squares and residual sum of squares each lose 1 d.f.

If more than one value is missing, a special modification of the above method has to be used and expert statistical advice should be sought.

8.4 Dealing With Non-Normal Variables

Everything that we have discussed so far under analysis of variance (A. of V.) has presupposed that we are dealing with a normal, or approximately normal, variable. Fortunately, many experimental variables do fall into this category, although it is probably not uncommon for investigators to blithely apply A. of V. on the *assumption* of normality and without bothering to make further checks. According to statisticians (e.g. Campbell, 1974), the A. of V. procedure is 'robust towards minor departures from normality'. But how robust is 'robust', and what is a 'minor' departure? These are questions for which answers are not easy to find.

If we suspect (or know) that our variable is *not* normal, then we have the choice of either trying to find a normalizing transformation, or we should switch to a non-parametric method of A. of V.

In addition to the problems of normality, and departures therefrom, there is a quite separate problem of *heteroscedasticity*. This technical term simply means that the within-group variances instead of being homogeneous (or just fluctuating randomly) show a steady increase as the variable itself takes on larger values. A common example of a heteroscedastic variable is provided by the Poisson distribution, where the variance is equal to the count. Thus, as the count gets larger, so automatically does the variance, and you therefore should not do A. of V. directly on Poisson data.

In finding a transformation the following points must be considered:

(1) normalization;

(2) maintenance (or acquisition) of *homo*scedasticity (the opposite of *hetero*scedasticity);

(3) there may also be an additional (and probably insurmountable) difficulty that some transformations which achieve normalization and *homo*scedasticity create spurious results in the interaction features of the A. of V. Obviously you would need expert help in this area.

8.4.1 Transform if possible

Unless the data are very extensive, e.g. of the order of 50 or more replicates at each treatment, it will probably be difficult to determine the appropriate transformation directly. So, unless you have such data—in which case, go and consult a statistician—it would probably be best to use transformations which other investigators have found to be satisfactory in systems analogous to your own. Meanwhile let us consider *measurements, proportions, counts* and *ranks*— the main categories into which statistical data fall.

(1) *Measurements.* Many measurement data can probably be handled by A. of V. without any preliminary transformation. However, there are some types of variable which are known to be log-normally distributed and they should therefore be subjected to a $\log_{10}$ transformation before processing through the A. of V. procedure. Examples of such data are serum insulin levels in man and mouse, and levels of various antibodies in animals (see Section 2.4.5).

(2) *Proportions.* These will definitely have to be transformed before A. of V., but you will probably need expert advice on which transformation is best for your particular system. If the variable follows the binomial distribution, then you may find the *angle* (also known as the *arcsin*) *transformation* to be appropriate. This converts a binomial proportion a/n into an angle θ defined by the relation $a/n = \sin^2\theta$. For example, the proportion 6/10 is converted as follows:

$$6/10 = 0.6000$$
$$\sqrt{0.6000} = 0.7746$$

The angle whose sine is 0.7746 is 50.8° and may be obtained either from trigonometric tables or from a pocket calculator. An example of a system for which the angle transformation is appropriate is the mouse-convulsion assay for insulin. See also Snedecor and Cochran (1967, p. 327).

However, the angle transformation is only one of several possibilities for proportion data. Others are the probit, logit, Cauchy, Wilson–Worcester and rectangular transformations. Chapter 17 of Finney's (1971) *Statistical Methods in Biological Assay* discusses these transformations and provides examples. Since this is too specialized a subject for discussion here, you are advised to seek expert help.

(3) *Counts.* With a Poisson variable, a commonly-used transformation is to take the square root of each observation and perform the A. of V. on the transformed values. With non-Poisson count data (see Chapter 6), either expert advice should be obtained or the data analysed non-parametrically (see below). In an analysis of variance of colony counts of heterotrophic bacteria in seawater samples, Ashby and Rhodes-Roberts (1976) used a $\log_e$ transformation. Snedecor and Cochran (1967, p. 329) given an example of A. of V. on plankton counts in net hauls. Here a $\log_{10}$ transformation was used before A. of V. was done.

(4) *Ranks.* These have to be analysed non-parametrically.

8.4.2 If not, go non-parametric

The main message to emerge from what follows is that the so-called non-parametric 'analysis of variance' is a very poor substitute for the 'real' A. of V. Notably it is much less versatile and is incapable of doing any more than the equivalent of single-factor A. of V. Being non-parametric, it has nothing to do with *variances*, but is concerned with the comparison of *medians*—in the same way as A. of V. (with a normal variable) is concerned with comparisons of *means*. The Friedman test, which we shall now do, is applicable to a randomized block design with equal numbers of results in each treatment. There is another procedure, the Kruskal–Wallis test, which is applicable to fully-randomized designs and can accept groups of unequal size. This test will not be presented here, but may be found in Campbell (1974).

Let us apply the Friedman test to the 24 serum-glucose results which are set out as a randomized block experiment in Table 8.6. In other words, we are processing the results as if we were unsure of their normal status and therefore treating them non-parametrically.

Table 8.9 Ranking of the serum glucose results of Table 8.6 for the Friedman test

| Block | Rank of treatment | | | |
	A	B	C	D
I	3	1	4	2
II	3	1	4	2
III	3	1	4	2
IV	3	2	4	1
V	3	1	4	2
VI	3	1	4	2
Treatment total (R)	18	7	24	11

Step 1. The first step of the Friedman test is to replace each observation in the table by its rank within its block, the smallest value being given rank 1. This has been done in Table 8.9. Note that ties are dealt with as in Section 4.4.3, e.g. if the values in Block IV had been 180, 139, 273, 139, the corresponding ranks would have been 3, 1½, 4, 1½.

Step 2. Add up the rank numbers in each treatment column to get an R-value (sum of ranks for each treatment).

Step 3. Calculate ΣR, the sum of the R-values:

$$\Sigma R = 18 + 7 + 24 + 11 = 60$$

Step 4. Make the check of arithmetic provided by:

$$\Sigma R = \tfrac{1}{2} m \cdot n \cdot (n+1) \dots\dots\dots\dots\dots\dots\text{Eq. 8.4}$$

where m = number of blocks = 6, n = number of treatments = 4,
$60 = \frac{1}{2} \times 6 \times 4(4+1)$, which checks.

Step 5. Calculate the test statistic:

$$S = \Sigma R^2 - (\Sigma R)^2/n \quad \ldots\ldots\ldots\ldots\ldots\ldots\ldots\ldots\text{Eq. 8.5}$$
$$= 18^2 + 7^2 + 24^2 + 11^2 - 60^2/4$$
$$= 170$$

Step 6. Consult the tabulated values of the Friedman statistic (Appendix A16) for the appropriate values of m and n. For $m = 6$ and $n = 4$ we find:

P	5%	1%	0.1%
S	76	100	128

If the found value of S is *larger* than the tabulated S at the $P = 5\%$, 1% or 0.1%, the differences in medians are, respectively, significant, highly significant and very highly significant. In our case the result is very highly significant (in agreement with the regular A. of V. procedure).

You will note that the Friedman table only deals with up to $n = 5$ treatment groups and up to $m = 10$ blocks. For larger experiments an alternative test statistic is calculated:

$$\chi^2 = \frac{12S}{mn\,(n+1)} \quad \ldots\ldots\ldots\ldots\ldots\ldots\ldots\ldots\ldots\text{Eq. 8.6}$$

where S is computed as in Eq. 8.5. One then determines the significance of the statistic by comparison with the tabulated values of χ^2 for $(n-1)$ degrees of freedom.

The main restriction on the Friedman test is that it cannot be elaborated into procedures analogous to the two-way A. of V., with infection, drug, interaction, etc. which we were able to employ when treating the data as being normally distributed.

The take-home message of this section is that there are great advantages in having a normally-distributed variable.

8.5 The Multi-Factorial Approach to Experiments

8.5.1 Some thoughts on strategy

Traditionally the scientific approach to experimentation consists in the manipulation of one variable at a time, while other factors are kept constant. The different variables are examined individually and methodically in this way, and data are accumulated and summarized as laws, rules or principles. This systematic, *unifactorial* approach is particularly important in fundamental research, especially during the initial probing of complex systems where the questions being asked have to be simple, and have to be asked one at a time.

The unifactorial strategy, although having a certain aesthetic purity, is not immune from criticism. For example, if only one variable at a time is manipulated, there may be no opportunity to observe possible *interaction* effects of the type we saw in the pertussis/adrenaline experiment. A further important point is that the unifactorial approach is relatively extravagant in consumption of resources, compared with the *multifactorial* approach which we shall now consider.

8.5.2 The economy of multifactorial designs

Particularly in agricultural and industrial investigations, the multifactorial approach to experimentation is firmly established. In contrast, it has received comparatively little attention from those engaged in fundamental biological research. However, for a recent example see Fannin et al., 1981 (*App. Env. Micro.*, **42**, 936–943). The essence of this approach is that several — 2, 3, 4 or more — variables are manipulated simultaneously within the same experiment and then afterwards their separate effects, and also their interactions with each other, are unravelled by a cunningly designed analysis of variance. There are two particularly useful features of multifactorial experiments:

(1) They are more economical of experimental material and investigator's time than unifactorial designs. Furthermore, the proportionate economy progressively increases as the number of variables gets larger.

(2) They permit the observation and quantification of interactions between variables which simply would not be observed in a unifactorial experiment.

Factorial designs are particularly important in agricultural research where experiments tend to be both lengthy and costly and where efficient design is therefore at a premium. Take a simple case of the effect of fertilizers on crop growth. It is well known that nitrogen (N), potassium (K) and phosphorus (P) are essential plant nutrients. The unifactorial approach to studying them would consist in, say, varying the N while keeping the K and P constant, and so on. In contrast, the multifactorial design can test 2 or 3 levels each of N, K and P simultaneously in the same experiment and then unravel their separate effects and also their interactions with each other. A microbiological equivalent of a multifactorial fertilizer experiment would be the formulation of a culture medium for optimum microbial growth or metabolite synthesis, or manipulation of environmental factors that affect biodegradation. The different factors might be carbon source, nitrogen source, vitamins and trace metals, each being tested at 2 or 3 levels and in all possible combinations.

Industrial processes are also amenable to multifactorial experimentation since anything up to a dozen or more variables may affect the final quality of a manufactured product, and for cost reasons it would be important to determine the optimum combinations.

With this as background, let us now take an example to see how the 'factorial' approach works.

8.5.3 A simple example

The simplest type of multifactorial design is the 2×2, or 2^2, factorial, which means we have 2 independent variables, or *factors*, each being tested at two levels. The two levels could be two different concentrations, or simply total absence of the factor at the low level and its presence at a particular concentration at the high level. If we had 3 variables each at 2 levels, this would be a 2^3 factorial, and so on. On the other hand, a 3^2 factorial design would have 2 variables, each at 3 levels. A complex design may have as many as 7 variables, each at 3 levels, giving a 3^7 factorial, or 2,187 treatment combinations. This gives some idea of the potential of factorial designs. But let us start with the 2^2.

We have, in fact, just dealt with a 2^2 factorial experiment, namely the pertussis/adrenaline experiment in mice. Let us now apply the factorial approach to it—it will give exactly the same results as we have already obtained—but it will get them by a different method and will introduce a new terminology which has the versatility of being usable with much more complex designs. The central feature of the analytical method and the terminology is the use of quantities known as *orthogonal contrasts* and which are represented by the symbol L with a subscript. It is not necessary to understand exactly what the L-values are, or why they are called orthogonal contrasts; all that we need to know is how to manipulate them correctly.

So let us refer back (Table 8.6) to the pertussis/adrenaline data as presented in the randomized block mode (the normal method of accumulating results in a factorial experiment).

Before starting the analysis of variance (A. of V.) there is some neat symbolism to be introduced to describe which level of each variable was applied in the various treatments. You will recall that the pertussis variable was applied at 2 levels which we could represent as 0 and + (uninfected and infected). Likewise, the adrenaline variable was applied at levels which we could also represent as 0 and + (saline and adrenaline). To avoid ambiguity with the +'s we represent the 2 levels of the pertussis variable as 0 and p, and the 2 levels of the adrenaline variable as 0 and a. We can then describe the 4 treatments by the scheme set

Table 8.10   Coding of treatments for the factorial approach to the analysis of the pertussis/adrenaline data

Original code letter for the treatment	A	B	C	D
Status of the pertussis variable in that group (p = pertussis-infected; 0 = uninfected)	O	p	O	p
Status of the adrenaline variable in that group (a = adrenaline; O = saline)	O	O	a	a
New code designation for the treatment	O	p	a	pa
Treatment total (T)	1,245	728	1,774	928

out in Table 8.10 which also lists the treatment totals (*T*-values). Note that to keep the new code designations as brief as possible, a combination like pertussis +, adrenaline O, is represented simply as p and not as pO, since the O is just bland diluent. We therefore, in our new coding, designate the treatments A, B, C and D as O, p, a and pa.

Now we are ready to use the *L*-terms whose main purpose is to subdivide the 'treatments' sum of squares (s.s.) in the analysis of variance. Previously this subdivision was done with an interaction table (Section 8.2.2) and the formulae in Table 8.3. This procedure gave the following s.s. (copied from Table 8.3):

$$\begin{aligned} &\text{pertussis:} &&77,407.042 \\ &\text{adrenaline:} &&22,143.375 \\ &\text{interaction:} &&4,510.04 \end{aligned}$$

We shall now get exactly the same results from the *L*-quantities ('contrasts') as follows.

Table 8.11 Use of contrasts and their multipliers to subdivide the treatments sum of squares

Contrast symbol	Multiplier to be used with treatment total (*T*)				Numerical value of *L* (contrast)	$(L)^2/4n$ (sum of squares)
	O	p	a	pa		
L_p	−1	1	−1	1	−1,363	77,407.04
L_a	−1	−1	1	1	729	22,143.375
L_{pa}	1	−1	−1	1	−329	4510.04

Referring to Table 8.11 we have three *L*-quantities:

(1) L_p = 'The contrast associated with the pertussis variable'.
(2) L_a = 'The contrast associated with the adrenaline variable'.
(3) L_{pa} = 'The contrast associated with pa interaction'.

Each of these *L*-values is a sum obtained by multiplying each *T*-value by the multipliers (known as coefficients of orthogonal contrasts) listed in the body of Table 8.11 and then adding up the 4 results, i.e.:

$$\begin{aligned} L_p &= 1245(-1) + 728(1) + 1774(-1) + 928(1) \\ &= -1245 + 728 - 1774 + 928 \\ &= -1363 \text{ as listed in the 3rd main column.} \end{aligned}$$

Similarly,

$$\begin{aligned} L_{pa} &= 1245 - 728 - 1774 + 928 \\ &= -329 \end{aligned}$$

We then calculate $(L)^2/4n$ for each *L*-value, where $n =$ the number of observations in each treatment group ($= 6$). This gives the results listed in the right-hand column of Table 8.11 which are exactly the same as the s.s. for pertussis, adrenaline and interaction as calculated previously.

So the use of the contrasts and multipliers allowed us to subdivide the treatments s.s. in the A. of V. and get the same results as the previously-used formulae. While the L-value method for a 2×2 factorial system is not necessarily simpler than what was used before, it is highly advantageous with the more complicated factorial designs.

The rest of the A. of V. procedure is the same as set out in Table 8.7, as are the F-tests and interpretations of the null hypotheses. So, to repeat: the main application of the L-values is to permit easy subdivision of the treatments s.s. in the analysis of variance.

Another use of the L-values is to provide measures of the effect of each factor in the experiment. Take the pertussis for example. We have, from Table 8.11, $L_p = -1363$. If we divide this by $2n$, where $n =$ the number of replications ($= 6$), we get what is called *the factorial-effect mean* of pertussis (on serum glucose) as $-1363/12 = -113.6$. This is to be interpreted as indicating that pertussis had the effect of *lowering* (because of the negative sign) the serum-glucose values by an average of 113.6 mg/100 ml.

Similarly for adrenaline as a variable, we have $L_a/2n = 729/12 = +60.7$. This indicates that adrenaline had the effect (averaged over infected and uninfected mice) of *increasing* (because of the positive sign) the serum glucose by 60.7 mg/100 ml.

The interaction term L_{pa} divided by $2n$ gives $329/12 = -27.4$. This means that the average net effect of applying both pertussis and adrenaline to a mouse was to give a serum-glucose concentration 27.4 mg/100 ml lower than would have been expected from simple addition of the separate (oppositely-acting) effect of pertussis alone or adrenaline alone. In other words, the -27.4 is a measure of the attenuation, caused by pertussis infection, of the adrenaline-mediated hyperglycaemia in the mouse.

These factorial-effect means, $L_p/2n$, $L_a/2n$ and $L_{pa}/2n$, thus provide useful arithmetic expressions for summarizing the separate effects of the main variables and also their interaction with each other.

9

Correlation, Regression and Line-Fitting Through Graph Points: Standard Curves

It is vain to do with more what can be done with fewer.

William of Occam (early 14th Century)

9.1 Preliminary Signposting

In England and Wales during the 1950s the Medical Research Council made a series of trials (1951, 1956, 1959) to assess the protective immunizing activity of the various whooping cough (pertussis) vaccines then in use. Each of 25 batches of vaccine was tested in groups of children to see how well it protected against an exposure to whooping cough in the home. This is expressed quantitatively (Table 9.1) as an 'immunity percentage in home exposures' (IPHE-value). Each vaccine was also tested in the laboratory for ability to immunize mice against experimental infection with the whooping cough bacillus, *Bordetella pertussis*. In Table 9.1 the vaccines are arranged down the page in order of

Table 9.1   Results[a] from the Medical Research Council trials on pertussis vaccines, and values of P, Q and $P-Q$ for calculating the Kendall coefficient of rank correlation (K)

Vaccine no.	Mice[b] (MP value)	Children[c] (IPHE value)	P	Q	$(P-Q)$
V2	0.496	13	24	0	24
G174	1.056	70	14	9	5
V1	1.171	22	22	0	22
V5	1.525	38	20	1	19
V3	1.665	39	19	1	18
V5a	1.841	39.5	18	1	17
V6	1.905	43	17	1	16
V3b	1.955	46	15	2	13
G61	2.00	70.5	13	3	10
V7	2.012	44	14	1	13
V4	2.030	26	14	0	14
087	2.300	78	10	3	7
V10	2.582	71	11	1	10
D231	2.594	93	1	10	−9
V8	2.851	86	7	3	4
V12a	2.857	76	8	1	7
V14	2.898	52	8	0	8
V16	2.899	84	7	0	7
V12b	2.928	91	3	3	0
V9	2.958	87	5	0	5
V17	2.983	92	1	3	−2
V11	3.023	96	0	3	−3
V15	3.090	91.5	0	2	−2
V19	3.157	90	0	1	−1
V20	3.183	89	—	—	—

[a]Adapted from Table VII in Medical Research Council (1959). Vaccination against whooping cough. *Brit. Med. J.* (i) 994–1000. *Reproduced by permission of the British Medical Association.*
[b]The figures are $\log_{10}$ of the relative potency, measured against vaccine G61 in the intracerebral mouse protection test, with 2.0 added to eliminate negative values.
[c]The figures are the percent of children who remained healthy after having been exposed to whooping cough in the home (IPHE = immunity percentage in home exposures).

increasing mouse-protective (MP)-value and with the corresponding IPHE-value given alongside.

These results are to be regarded as a convenient working example of a general category of data where there are 2 types of measurements, i.e. 2 variables, associated with each experimental object or unit. Such data allow exploration of the following questions:

(1) Are the 2 measured variables *correlated?* That is, does a particular value of one variable tend to go hand-in-hand with a particular value of the other? And if so, how close is this hand-in-hand association?

(2) If there is some kind of relationship between the 2 variables, can a straight line be fitted through it and can such a graph be used to predict the value of one variable, given a value of the other?

These questions lead into a consideration of the statistical procedures known as *correlation analysis* and *regression analysis*. As we shall see, there is considerable overlap of methods and underlying assumptions between correlation and regression, but also some points of sharp distinction. That is to say, there are some data which are suitable for regression analysis but not for correlation analysis, and vice versa.

In the example chosen, we are dealing with only 2 variables, MP and IPHE. But sometimes there are many variables, e.g. the growth or life-span of an organism (animal, plant or microbe) may be affected by a wide variety of physical and chemical influences on which we have measurements, and we may wish to do a preliminary sorting out to establish *which* of the several measured variables affect the organisms and which do not. Here correlation analysis will be useful. Then having picked out the influential variables, we can use regression analysis to study the *shape* of the relationship between that variable and growth, e.g. is it linear or curved? Regression analysis is thus applicable for dose–response relationships: i.e. when we set up a series of known doses or concentrations of some agent and then measure the corresponding responses so as to produce a standard curve or graph for the system. Let us now return to the pertussis vaccine data.

9.2 Correlation

The first step in correlation analysis is to decide whether the 2 variables are normally distributed. If we have reason to believe that they *are*, then we can calculate the 'product–moment correlation coefficient' (r). On the other hand, the evidence of normality may be unconvincing or lacking, in which case we should use the more cautious non-parametric approach, as we did with the tests of significance in Chapter 4. The corresponding non-parametric procedure yields the 'Kendall rank-correlation coefficient' (K). For exercise purposes, let us apply both procedures to the same set of data, justifying our actions by different sets of underlying assumptions.

9.2.1   Kendall coefficient of rank correlation

Our reason for using the Kendall approach is that we have no basis for regarding either MP or IPHE as being normally distributed. Therefore a non-parametric analysis avoids us having to make what may be an unjustified assumption. The steps are as follows:

(1) Plot a graph of the 2 variables as done in Fig. 9.1. It is always useful to see by naked eye if the pattern of points shows some kind of trend or whether they are quite randomly scattered on the page. Remember that the statistical calculation may do little more than confirm in a quantitative fashion what the eye perceives subjectively.

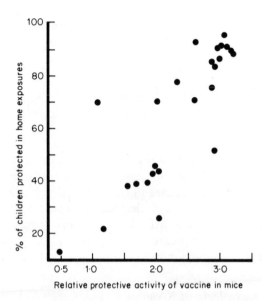

Fig. 9.1   Protective activity of pertussis vaccines in human infants and in mice: a correlation diagram in which each point represents a different batch of vaccine (plotted form the data in Table 9.1)

(2) Arrange the order of the abscissa variable — here MP has been assigned rather arbitrarily to the abscissa — in *increasing rank order* down the page. This has already been done in Table 9.1. Tabulate the values of the other variable (IPHE) alongside, as in column 3 of that table.

(3) The essence of the Kendall procedure is to obtain the quantities P, Q and $P - Q$, as in columns 4, 5 and 6 of the table.

(4) To get P, take each IPHE-value starting with the topmost and count the number of *larger* values that lie in the column *below* it. The first entry of 13 has 24 larger values, so we enter $P = 24$ in the adjacent column. Similarly, the next IPHE-value of 70 has 14 larger values located spatially underneath it in the column. Go down the column in this way, entering each P value in turn.

(5) To get Q, proceed similarly, except that with each IPHE-value, count the number of *smaller* IPHE-values that lie below it.

(6) Calculate $(P-Q)$ for each (P,Q) pair.

(7) Add up the $(P-Q)$-values to get:

$$S = \Sigma(P-Q) = 219 - 17 = 202$$

(8) Calculate the Kendall rank-correlation coefficient (K) by the formula:

$$K = \frac{2S}{N(N-1)} \qquad \dots\dots\dots\dots\dots\dots\dots\dots Eq.\ 9.1$$

where $N =$ the number of pairs of IPHE- and MP-values. In this example $N = 25$. Therefore

$$K = \frac{2(202)}{25(24)}$$

$$K = 0.673$$

(9) Interpretation: A positive value of K, as here, tells us that the 2 variables are positively correlated, i.e. as one gets larger, the other tends to get larger also. A K-value of $+1.0$ would mean perfect positive correlation; to get this would require that the rank-order of the vaccines would be the same whether they were arranged in increasing MP-value or in increasing IPHE-value. The found K-value of 0.673 is a measure of the 'purity' of the correlation and indicates that there is considerable departure from perfection.

9.2.2 Pictorial equivalents of K-values

Fig. 9.2 illustrates data similar to those just considered but with K-values of $+1.0$, -1.0 and 0.0. Note that to get $+1.0$ or -1.0 does not require the points to fall on a straight line; all that is needed is a steady upward (for $+1.0$), or downward (for -1.0), trend in the y-variable for each increment in the x-variable. Note also that $K = 0$, or close to 0, can arise in several ways: by random scatter (9.2c) by horizontal alignment (9.2d) and by an inverted U-shaped or arch-shaped distribution (9.2e).

9.2.3 Statistical significance of K

Just to calculate the correlation coefficient K is not the end of the job: the next step is to determine whether the found value of K is significantly different from 0. Otherwise we cannot exclude the possibility that the apparent association between the two variables has arisen by chance. The procedure for investigating

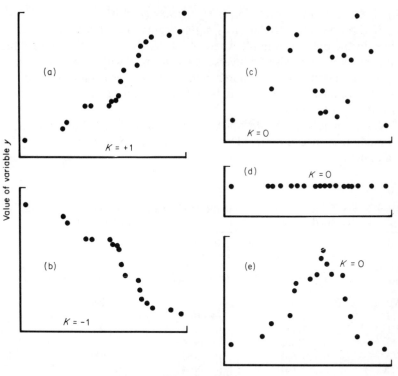

Fig. 9.2   Samples of data that exhibit various patterns of correlation, and showing how a Kendall coefficient of correlation (K) equal to zero can arise in several different ways

significance varies according to whether the number of points (N) is greater or less than 10.

For $N>10$, as here, substitute S and N into:

$$Z = \frac{S\sqrt{18}}{\sqrt{[N(N-1)(2N+5)]}} \quad \ldots\ldots\ldots\ldots\ldots \text{Eq. 9.2}$$

With $S=202$ and $N=25$, this gives:

$$Z = \frac{202\sqrt{18}}{\sqrt{[25(25-1)(2\times25+5)]}}$$

$$= 4.72$$

The rules for interpreting the found value of Z are as follows:
  (a) if $Z<1.96$, the correlation is *not* significant ($P>5\%$);

(b) if $Z \geqslant 1.96$, the correlation is *significant* $(P \leqslant 5\%)$;

(c) if $Z \geqslant 2.58$, the correlation is *highly* significant $(P \leqslant 1\%)$.

Thus, in the present example where $Z = 4.72$, the correlation is very highly significant. Note that with an inverse correlation, S will have a negative sign, as will Z. This negative sign is ignored in making the interpretation of significance level. The technical name for Z is 'standardized normal deviate' and is dealt with further in the Glossary.

If N is 10 or less, Z should not be calculated by Eq. 9.2 and recourse has to be made to a special table. (This is not reproduced here but may be found in Campbell, 1974, p. 355.)

Table 9.2   As Table 9.1, except that the IPHE values are the original data,[a] which included some 'tied' results (shown here linked with arrows)

| Vaccine no. | Protective activity in | | P | Q | $(P-Q)$ |
	Mice[b] (MP value)	Children[c] (IPHE value)			
V2	0.496	13	24	0	24
G174	1.056	→70	13	9	4
V1	1.171	22	22	0	22
V5	1.525	39◄	18	1	17
V3	1.665	39◄	18	1	17
V5a	1.841	39◄	18	1	17
V6	1.905	43	17	1	16
V3b	1.955	46	15	2	13
G61	2.00	→70	13	3	10
V7	2.012	44	14	1	13
V4	2.030	26	14	0	14
087	2.300	78	10	3	7
V10	2.582	71	11	1	10
D231	2.594	93	1	10	-9
V8	2.851	86	7	3	4
V12a	2.857	76	8	1	7
V14	2.898	52	8	0	8
V16	2.899	84	6	0	6
V12b	2.928	91◄	2	3	-1
V9	2.958	87	5	0	5
V17	2.983	92	1	3	-2
V11	3.023	96	0	3	-3
V15	3.090	91◄	0	2	-2
V19	3.157	90	0	1	-1
V20	3.183	89	—	—	—

[a]Adapted from Table VII in Medical Research Council (1959). Vaccination against whooping cough. *Brit. Med. J.* (i) 994–1000. *Reproduced by permission of the British Medical Association.*

[b]The figures are $\log_{10}$ of the relative potency, measured against vaccine G61 in the intracerebral mouse protection test, with 2.0 added to eliminate negative values.

[c]The figures are the percent of children who remained healthy after having been exposed to whooping cough in the home (IPHE = immunity percentage in home exposures).

9.2.4 Dealing with 'ties'

In any operation involving the rank-ordering of data, one can expect to find values that are numerically identical, i.e. are 'tied'. The Kendall procedure, like other non-parametric methods, requires some modification to deal with tied values, and to illustrate what is needed let us take the data of Table 9.2 as our working example. These are based on the original MRC results and contain three sets of tied values in the IPHE column, as shown by the arrowed brackets. The calculation of P and Q is done basically as before but with this restriction: If the particular MP/IPHE pair being dealt with is tied (in either its MP- or its IPHE-value, or both) with any other pair *located below it* in the table, then these other pairs are *excluded* from the counts for P and Q of the MP/IPHE pair in question.

In order to do this job without making mistakes, it is useful to have a supply of strips of tracing paper which can be laid over the MP and IPHE columns whenever a tie is encountered and the appropriate values scratched out.

Starting at the top of the IPHE column in Table 9.2, vaccine V2 has $P = 24$ and $Q = 0$ as before. However, the next one, G174, is tied with G61. Therefore lay the tracing paper on the IPHE column and temporarily scratch out G61. This gives $P = 13$ and $Q = 9$ for G174. Proceed in this way and apply the same treatment to the 2 other sets of tied results. From the values of P and Q in Table 9.2, we get $\Sigma(P-Q) = 196 = S$.

Before going on to calculate the Kendall coefficient (K), we need quantities A and D which act as correction factors for the tying in the abscissa (A) and ordinate (D) columns, respectively. A is defined by:

$$A = \tfrac{1}{2}\Sigma J(J-1) \qquad \dots\dots\dots\dots\dots\dots\dots\dots\dots\dots\dots\text{Eq. 9.3}$$

where J is the number of components in a tie in the abscissa (MP) column. In this example we have no ties in the MP column and therefore $J = 0$ and $A = 0$.

However, in the IPHE column we have ties of 2, 3 and 2 values. Analogous to Eq. 9.3 is:

$$D = \tfrac{1}{2}\Sigma H(H-1) \qquad \dots\dots\dots\dots\dots\dots\dots\dots\dots\dots\dots\text{Eq. 9.4}$$

where H is the number of components in a tie in the ordinate (IPHE) column. This gives:

$$
\begin{aligned}
D &= \tfrac{1}{2}\,[2(2-1)+3(3-1)+2(2-1)] \\
&= \tfrac{1}{2}\,[2+6+2] \\
&= 5
\end{aligned}
$$

The Kendall coefficient (K) is then defined by:

$$K = S/\sqrt{[\tfrac{1}{2}N(N-1)-A]\cdot[\tfrac{1}{2}N(N-1)-D]} \qquad \dots\dots\text{Eq. 9.5}$$

If there are no ties, both A and D are zero and the equation reduces to Eq. 9.1. Substituting our values $S = 196$, $N = 25$, $A = 0$ and $D = 5$ gives:

$$K = 196/\sqrt{[\,\tfrac{1}{2} \times 25 \times 24 - 0\,] \cdot [\,\tfrac{1}{2} \times 25 \times 24 - 5\,]}$$

$$= 0.659$$

which is only slightly lower than the value found previously, from the data with no ties.

9.2.5   Correlation with normally-distributed variables

The Kendall coefficient of rank correlation (K) is a non-parametric quantity based on ranking the values of the variables and without assumptions about whether either of them is normally distributed. An alternative procedure is to calculate the 'product-moment correlation coefficient' (r). For this, we make the assumption that *both* variables are normally distributed. Fig. 9.3 presents

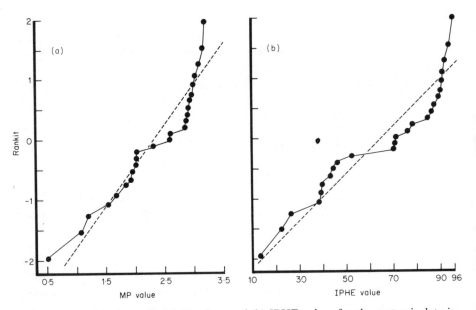

Fig. 9.3   Rankit plots of (a) MP-values and (b) IPHE-values for the pertussis data in Table 9.1. Note that both show the 'chair' shape characteristic of a distribution with negative kurtosis

the rankit plots of MP and of IPHE and suggests that this assumption is acceptable, although the tendency for a 'chair' shape suggests that there may be some degree of kurtosis.

With two variables, x and y, the formula for the product–moment correlation coefficient (r) is:

$$r = \frac{\Sigma(x-\bar{x})(y-\bar{y})}{\sqrt{\Sigma(x-\bar{x})^2 \cdot \Sigma(y-\bar{y})^2}} \quad \dots \dots \dots \dots \dots \text{Eq. 9.6}$$

where we can take x as representing MP and y as IPHE. Taking the bottom line of the equation first, and using the algebraic identity in Section 1.7.4, we have:

$$\begin{aligned}
\Sigma(x-\bar{x})^2 &= \Sigma x^2 - (\Sigma x)^2/N \quad \dots \dots \dots \dots \dots \dots \text{Eq. 9.7} \\
&= 0.496^2 + 1.056^2 + \dots + 3.183^2 - 57.959^2/25 \\
&= 147.624437 - 134.369827 \\
&= 13.25461
\end{aligned}$$

where the x-values were taken from column 2 of Table 9.1, and $N = 25$. Similarly:

$$\begin{aligned}
\Sigma(y-\bar{y})^2 &= \Sigma y^2 - (\Sigma y)^2/N \quad \dots \dots \dots \dots \dots \dots \text{Eq. 9.8} \\
&= 13^2 + 70^2 + \dots 89^2 - 1627^2/25 \\
&= 122{,}131 - 105{,}885.16 \\
&= 16{,}245.84
\end{aligned}$$

The top line of Eq. 9.6 is calculated with an analogous algebraic identity:

$$\begin{aligned}
\Sigma(x-\bar{x})(y-\bar{y}) &= \Sigma xy - (\Sigma x)(\Sigma y)/N \quad \dots \dots \dots \dots \dots \text{Eq. 9.9} \\
&= (0.496 \times 13) + \dots + (3.183 \times 89) - 57.959 \times 1627/25 \\
&= 4153.031 - 3771.97172 \\
&= 381.05928
\end{aligned}$$

We therefore have:

$$\begin{aligned}
r &= \frac{381.05928}{\sqrt{13.25461 \times 16{,}245.84}} \\
&= 0.821
\end{aligned}$$

As with the Kendal rank-order correlation coefficient (K), the value of r indicates a strong positive correlation between the two variables.

To finish off the calculations, we determine the statistical significance of the found value of r. Thus we set up a null hypothesis which states that 'there is *no* association between x and y (i.e. no tendency for one to go hand-in-hand with the other) and the apparent correlation observed could have arisen through random-sampling fluctuations'. The test of this hypothesis is to enter Appendix A17 with $N-2$ degrees of freedom, i.e. 23 degrees of freedom (d.f.) in this case, and read the tabulated values of r at the conventional levels of probability. Since only 20 and 25 d.f. are provided in the table, we can interpolate approximate values for 23 d.f. as follows:

if $r < 0.397$, the correlation is *not* significant at $P > 5\%$;
if $r \geqslant 0.397$, the correlation is *significant* at $P \leqslant 5\%$;
if $r \geqslant 0.507$, the correlation is significant at $P \leqslant 1\%$;
if $r \geqslant 0.619$, the correlation is significant at $P \leqslant 0.1\%$.
With this example, an r-value of 0.82 is very highly significant — even beyond the 0.1% probability level. The parametric and non-parametric investigations of the correlation of MP and IPHE are therefore in agreement.

9.2.6 The meaning of r

Like the Kendall coefficient (K), r can take on values only between -1 through 0 to $+1$ and, as before, a positive coefficient signifies a positive correlation. Also, as before, a value of 0 or close to 0 can arise in several different ways, namely if there is a completely random scatter of points, if the points lie on a straight line parallel to the x-axis, or if they lie on a U-shaped or inverted U-shaped curve. K and r do, however, differ in the arrangement of points that give a coefficient of 1.0. Thus, for $r = 1.0$, all the points must lie exactly on a straight line, whereas this is not required for $K = 1.0$, although K *does* equal 1.0 if the points *do* lie on a straight line.

Values of r numerically less than 1.0 imply some departure from linearity through the points being scattered. If the graphical plot shows that the points lie on a curve, then the product–moment correlation coefficient should *not* be used as a measure of the association between the 2 variables and expert advice should be sought.

9.3 Linear Regression — Or Fitting the Best Straight Line Through a Series of Points

So far in this chapter we have avoided drawing a line through the scattered points of the MRC trial. This was deliberate, because correlation is not a line-drawing exercise but simply a procedure for determining *the degree of association* between 2 variables and expressing this on a dimensionless scale from -1 through 0 to $+1$. If now we wish to fit a line through the points, to represent the dependence of one variable upon another, then we become involved in *regression*.

Naturally, if you have a set of experimental points from some highly reproducible system like a spectrophotometric calibration graph in a biochemical analysis, then you may not need statistics, but just a ruler. Much the same applies with antibiotic assays on culture plates, where the zone diameter of growth inhibition is linearly related to the logarithm of the concentration of the diffusing substance (Section 10.2.1). For many purposes with such systems, a line can be fitted perfectly satisfactorily by eye and the graph used as a standard curve.

This situation does not, however, apply to such collections of data as those from the MRC whooping-cough trials, where the ruling of a straight line by eye alone is likely to be rather subjective, and this is where linear regression comes in.

9.3.1 Procedure for linear regression

First plot the points to see if they *look* as if they might be scattered around a straight line, or whether they seem to follow a curved relationship. If the latter, you should be cautious about going ahead with linear regression and should think instead of a transformation of one of the variables so as to achieve linearity. Otherwise all you would be doing is to fit a straight line through data which should really be represented by a curve (which would not be very useful).

Algebraically, a straight line is represented by an equation of the type:

$$y = a + bx \qquad \ldots\ldots\ldots\ldots\ldots\ldots \text{Eq. 9.10}$$

where a and b are constants, or *parameters*, of that particular system. The essence of linear regression is to calculate a and b, so that thereafter the position of the straight line is determined by b, the slope, and a, the intercept, on the y-axis when $x = 0$.

9.3.2 Calculating the regression coefficient (b)

The regression coefficient (b), is the same as the slope constant of the linear equation $y = a + bx$ and is calculated from the formula:

$$b = \frac{\Sigma(x - \bar{x})(y - \bar{y})}{\Sigma(x - \bar{x})^2} \qquad \ldots\ldots\ldots\ldots \text{Eq. 9.11}$$

Let us now substitute the MRC trial data into this equation, using the convention taken for correlation that we would treat the mouse potency (MP) of each vaccine as the independent variable (x) and the human potency (IPHE) as the dependent variable (y). With this agreed, we find that most of the arithmetic has already been done in calculating the product–moment correlation coefficient (r) in Section 9.2.5. Thus, from Section 9.2.5, we have:

$$\Sigma(x - \bar{x})(y - \bar{y}) = 381.05928$$

and

$$\Sigma(x - \bar{x})^2 = 13.25461$$

Therefore, from Eq. 9.11:

$$b = 381.05928/13.25461$$
$$= 28.7492$$

The value of b tells us that when we fit the straight line through the points of the MP/IPHE graph, a unit change in MP will be accompanied by a change of 28.7492 on the IPHE scale.

9.3.3 Fitting the straight line

In order to fit the straight line we still need the other constant a of Eq. 9.10 and we get it by using a variant of this equation, namely:

$$y = \bar{y} + b(x - \bar{x}) \qquad \dots\dots\dots\dots\dots\dots\dots\dots \text{Eq. 9.12}$$

From Section 9.2.5, we have:

$$\bar{y} = \Sigma y/N = 1627/25 = 65.08$$

$$\bar{x} = \Sigma x/N = 57.959/25 = 2.3184$$

Therefore, Eq. 9.12 yields:

$$y = 65.08 + 28.7492(x - 2.3184)$$

which simplifies to:

$$y = -1.572 + 28.7492x \qquad \dots\dots\dots\dots\dots\dots \text{Eq. 9.13}$$

This equation is the *regression of y on x*, i.e. it predicts the most probable value of y for each value of x. To fit the line on the graph we need to calculate two pairs of (x,y) points, e.g. if we put $x=0.5$ in Eq. 9.13, $y=12.80$; and $x=3.0$ gives $y=84.68$. In Fig. 9.4(a) these two points have been plotted, and the line between them drawn.

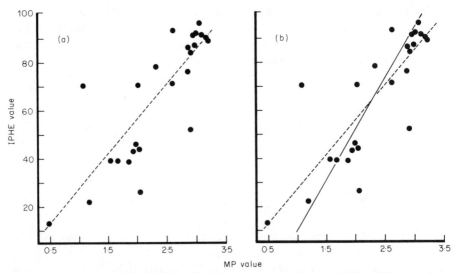

Fig. 9.4   Linear regression relationships in the pertussis vaccine trial data: (a) regression of IPHE on MP and (b) regression of IPHE on MP (dotted line as in (a)) with regression of MP on IPHE (solid line, superimposed)

9.3.4 The regression of x on y

The regression calculation, as just done, assumes that one of the variables, in this case x, has its values fixed with negligible error, so that *all* of the scatter of the experimental points away from the fitted line is due to random fluctuations in y. The straight line that was fitted in the preceding section was based on these assumptions and is referred to as the 'regression of y on x'.

Another regression line that could be fitted through the same set of points is described as the 'regression of x on y'. Here we make the converse assumption of believing that each y-value is known with negligible error and that all the fluctuations of points about the fitted straight line are due to the variability of x. The equations are similar to those used previously, but with adjustments to cater for the new assumptions. Thus, the new regression line can be represented by:

$$x = a' + b'y \qquad \ldots\ldots\ldots\ldots\ldots\text{Eq. 9.14}$$

where a' and b' are the new values for intercept and slope.

The regression coefficient (b') is defined by:

$$b' = \frac{\Sigma(x-\bar{x})(y-\bar{y})}{\Sigma(y-\bar{y})^2} \qquad \ldots\ldots\ldots\text{Eq. 9.15}$$

where all the quantities have already been calculated in Section 9.2.5 above, i.e.:

$$\Sigma(x-\bar{x})(y-\bar{y}) = 381.05928$$
$$\Sigma(y-\bar{y})^2 = 16{,}245.84$$
$$\therefore b' = 0.02346$$

To calculate the other constant (a') we use:

$$x = \bar{x} + b'(y-\bar{y}) \qquad \ldots\ldots\ldots\ldots\text{Eq. 9.16}$$

where $\bar{x} = 2.3184$ and $\bar{y} = 65.08$, from Section 9.3.3. Therefore:

$$x = 2.3184 + 0.023456(y - 65.08)$$
$$x = 0.7919 + 0.023456y \qquad \ldots\ldots\text{Eq. 9.17}$$

Fig. 9.4(b) shows this regression line superimposed on the previously obtained line for y on x.

One of the purposes of this exercise is to show that there are two 'best' straight lines which can be fitted through the data points. The calculation of the parameters of each line starts from different assumptions and highlights a recurrent message in this text, namely that the answer you get in a statistical calculation depends on what assumptions you start with.

Andrew Lang 1844–1912. He uses statistics as a drunken
man uses lamp-posts — for support rather than illumination.
*From MacKay, 1977, and reproduced through the courtesy of
the Institute of Physics*

9.3.5   *The original meaning of regression*

At first sight the word 'regression' may have little obvious relationship to the
way we have been using it to describe the association between two variables,
x and y. The dictionary definition is: 'The action of returning to or towards
a point of departure.' In this sense it is the opposite of *progression*. The term
was introduced into biostatistics by the pioneer mathematical geneticist Galton,
in 1889. He propounded a 'Law of Universal Regression' in which he stated
that 'each peculiarity in a man is shared by his kinsman, but *on the average
in a lesser degree*'. For example, in comparing the heights of fathers and sons,
it has been found that although tall fathers tend to have tall sons, *on average*
the heights of a group of sons from tall fathers is less than the average height
of the tall fathers themselves. There is thus a *regression*, or going back, of sons'
heights towards the average height of all men. This original, rather restricted,
usage of regression has expanded with the passage of time to the more general
meaning of 'a relation between 2 or more variables which can be expressed as
an algebraic function and where one or both variables are subject to variation'.
It is in this sense that we have used the word here.

9.4 Standard Curves and Interpolation of Unknown y-Values Thereon

A common requirement in biological work is to prepare a straight-line graph which shows the relationship between the amount of a substance and the effect, or response, which it produces in some experimental system. Responses produced by unknown samples can then be interpolated and read off in terms of the standard. Let us take as our example a biuret assay for protein (Herbert *et al.*, 1971) as illustrated by the set of duplicate results provided in Table 9.3. Here we have a series of paired optical density readings for different amounts of standard protein, together with the optical densities given by 3 unknown solutions whose protein concentration we wish to measure and which were also

Table 9.3 Biuret test: optical densities given by standard protein and by 3 unknown samples. The method of Herbert *et al.* (1971) was used

Sample	Amount used	O.D. of duplicate tests
Standard protein	1.0 mg	0.056, 0.080
	2.0 mg	0.122, 0.138
	4.0 mg	0.256, 0.278
	6.0 mg	0.400, 0.394
	8.0 mg	0.531, 0.543
Unknown A	1.0 ml	0.051, 0.041
Unknown B	1.0 ml	0.274, 0.286
Unknown C	1.0 ml	0.497, 0.506

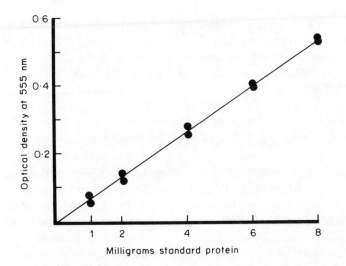

Fig. 9.5 Standard line for protein estimation by the Biuret method

set up in duplicate. From the standard graph (Fig. 9.5) we can read off the protein values for the unknowns as:

A (average O.D. = 0.046): 0.70 mg in 1.0 ml
B (average O.D. = 0.280): 4.1 mg in 1.0 ml
C (average O.D. = 0.501): 7.4 mg in 1.0 ml

A graphical evaluation of this kind is always useful as a first step in data processing. For example, it allows one to see if a straight line is a satisfactory representation of the dose–response relationship, in addition to permitting the insertion of the unknowns.

The main deficiency of the purely graphical method is that it does not go far enough and, in particular, does not provide any estimate of error. Also, the line fitted by eye may not be the best-fitting line, and unknown responses may be difficult to interpolate with proper accuracy. There is therefore good justification for evaluating the results algebraically, especially since routinely one would get a computer to do the number-crunching. Some useful procedures now follow.

9.4.1   Analysis of variance of the regression line

Although in the present case (Fig. 9.5) there would appear not to be any problem with deviations from linearity, this may not always be so, and it is useful therefore to apply the analysis of variance procedure as routine. Furthermore, the statistical quantities generated in this analysis are used to fit the best dose–response line, to estimate the strength of the unknowns and to calculate the 95% confidence limits. Therefore the effort is well worthwhile.

In Table 9.4 we have a re-tabulation of the protein assay data with extra columns for the response totals (T-values) and response means ($\bar{y}$-values). In order to distinguish beween the standard and the unknowns, a subscript notation is needed. It makes the formulae look complicated but they are basically the same as before.

In the lower half of Table 9.4 are summarized the various statistical quantities we shall need for the subsequent calculations. Note that these are derived from the standard data only. Thus, we have the following:

$$N_s = \text{number of data points for standard} = 10$$
$$\Sigma y_s = 0.056 + 0.080 + \ldots + 0.543 = 2.798$$

from which we get $\bar{y}_s = 0.2798$. Similarly, for the sum of squares (s.s.) of y_s, we have:

$$\Sigma y_s^2 = 0.056^2 + 0.080^2 + \ldots + 0.543^2 = 1.07833$$

Table 9.4 Retabulation of Biuret data, addition of algebraic symbols and calculation of basic statistics

Sample	Dose (mg) (x)	Responses (y)		Response total (T)	Response mean $(\bar{y})$
Standard	1	0.056	0.080	0.136	
	2	0.122	0.138	0.260	$\bar{y}_s = 0.2798$
	4	0.256	0.278	0.534	for all 10
	6	0.400	0.394	0.794	values
	8	0.531	0.543	1.074	
Unknown A	x_A	0.051	0.041	0.092	0.046
Unknown B	x_B	0.274	0.286	0.560	0.280
Unknown C	x_C	0.497	0.506	1.003	0.501

Statistics derived from standard data:

$N_s = 10$

$\Sigma y_s = 2.798$ $\qquad \bar{y}_s = 0.2798 \qquad (\Sigma y_s)^2/N_s = 0.7828804$

$\Sigma y_s^2 = 1.07833$ $\qquad \Sigma y_s^2 - (\Sigma y_s)^2/N_s = 0.2954496$

$\Sigma x_s = 42$ $\qquad \bar{x}_s = 4.2 \qquad (\Sigma x_s)^2/N_s = 176.4$

$\Sigma x_s^2 = 242$ $\qquad \Sigma x_s^2 - (\Sigma x_s)^2/N_s = 65.6$

$\Sigma x_s.y_s = 16.148$ $\qquad (\Sigma x_s)(\Sigma y_s)/N_s = 11.7516$

$\Sigma x_s.y_s - (\Sigma x_s)(\Sigma y_s)/N_s = 4.3964$

$\Sigma T_s^2/2 = 1.077582$ $\qquad \Sigma T_s^2/2 - (\Sigma y_s)^2/N_s = 0.2947016$

and the correction factor

$$(\Sigma y_s)^2/N_s = (2.798)^2/10 = 0.7828804$$

The corresponding x-values are obtained in a similar way:

$$\Sigma x_s = 2(1 + 2 + 4 + 6 + 8) = 42;$$

the 2 outside the parentheses is because each dose level has duplicate results. $\bar{x}_s$ is Σx_s divided by N_s and equals 4.2

$$(\Sigma x_s)^2/N_s = 42^2/10 = 176.4$$

The s.s. of the x's is:

$$\Sigma x_s^2 = 2(1^2 + 2^2 + \ldots + 8^2) = 242$$

The sum of products of x and y are also straightforward, but note that we can save some arithmetic by multiplying each x by its corresponding T-value, which represents the sum of the y's, i.e.:

$$\Sigma x_s \cdot y_s = (1 \times 0.136) + (2 \times 0.260) + \ldots + (8 \times 1.074)$$
$$= 16.148$$

Then $(\Sigma x_s)(\Sigma y_s)/N_s = 42 \times 2.798/10 = 11.7516$

We are now ready to do the analysis of variance of the regression, so let us set down the branching diagram for degrees of freedom:

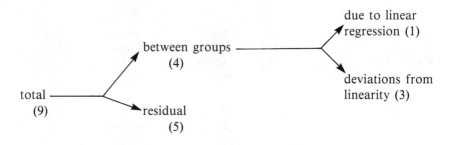

With $N_s = 10$ observations, we have a total of 9 d.f. With 5 groups, we have 4 d.f. for 'between groups' and 5 d.f. within groups or 'residual'.

Table 9.5 Analysis of variance of the linear regression of the Biuret protein standard. Most of the values in the sums of squares column are from the footnote to Table 9.4

Source of variation	Degrees of freedom	Sums of squares	V	F^a
Total	9	$\Sigma y_s^2 - (\Sigma y_s)^2/N_s$ $= 0.2954496 \ldots \ldots$ ①	—	—
Between groups	4	$\Sigma T_s^2/2 - (\Sigma y_s)^2/N_s$ $= 0.2947016 \ldots \ldots$ ②	—	—
Due to linear regression	1	$\dfrac{[\Sigma x_s y_s - (\Sigma x_s)(\Sigma y_s)/N_s]^2}{\Sigma x_s^2 - (\Sigma x_s)^2/N_s}$ $= (4.3964)^2/65.6$ $= 0.294639 \ldots \ldots$ ③	0.2946	1969.25 (**)
Deviations from linearity	3	② − ③ $= 0.000062$	0.000021	0.14[b] (NS)
Residual	5	① − ② $= 0.000748$	0.0001496	—

$s = \sqrt{\text{residual mean square}} = \sqrt{0.0001496} = 0.01223$.

[a]NS = not significant $(P > 5\%)$: tabulated F for 3,5 d.f. and $P = 5\% = 5.41$.
** = highly significant $(P \leqslant 1\%)$: tabulated F for 1,5 d.f. and $P = 1\% = 16.26$.
[b]As with other cases (e.g. Tables 7.1 and 8.5), where the found F is less than 1.0, the ratio has to be inverted for significance testing. This gives $1/0.14 = 7.14$ with 5 and 3 d.f. and is still NS, since the tabulated values are 9.01 and 28.24.

The 4 between-group d.f. can be divided into 2 components: 'due to linear regression' with 1 d.f., and 'deviations from linearity', with $4-1=3$ d.f. The 'due to linear regression' term always has just 1 d.f. because there is just one regression line involved, while the other term, 'deviations from linearity', has the remaining d.f. which reflect the various modes of kinking that could occur with the number of dose points in question.

The analysis of variance is set out in Table 9.5 and follows the pattern of the d.f. diagram. All the quantities needed for the s.s. have already been calculated in the footnotes of Table 9.4. The only unusual-looking term is that for the s.s. due to linear regression, but both numerator and denominator have already been calculated. The output from the analysis is the right-hand column with the F-ratios which show that we have a highly significant linear regression but no significant departure from linearity. The square root of the residual mean square gives us the standard deviation, $s = 0.01223$, for use in subsequent calculations.

Table 9.6 provides a generalized analysis of variance table for a linear regression with k doses, at each of which there are n replicate observations.

Table 9.6   Generalized analysis of variance of a linear regression with k doses, at each of which there are n replicate observations

Source of variation	Degrees of freedom	Sums of squares	V	F
Total	$nk-1$	$\Sigma y^2 - (\Sigma y)^2/nk$ ...... ①	—	—
Between doses	$k-1$	$\dfrac{\Sigma T^2}{n} - \dfrac{(\Sigma y)^2}{nk}$ ...... ②	—	—
Due to linear regression	1	$\dfrac{[\Sigma y \cdot x - (\Sigma y)(\Sigma x)/nk]^2}{\Sigma x^2 - (\Sigma x)^2/nk}$ .... ③	③	$\dfrac{③}{⑦}$
Deviations from linearity	$k-2$	② − ③ = ④	$\dfrac{④}{k-2} = ⑥$	$\dfrac{⑥}{⑦}$
Residual	$k(n-1)$	① − ② = ⑤	$\dfrac{⑤}{k(n-1)} = ⑦$	—

9.4.2   Parameters of the regression line

The regression line may be represented by:

$$y = \bar{y}_s + b(x - \bar{x}_s) \qquad \text{..........................Eq. 9.18}$$

for which most of the calculations have already been done above, i.e.

$$\bar{y}_s = 0.2798, \text{ from the footnote of Table 9.4}$$

$$b = \frac{\Sigma x_s \cdot y_s - (\Sigma x_s)(\Sigma y_s)/N_s}{\Sigma x_s^2 - (\Sigma x_s)^2/N_s} \quad \ldots\ldots\ldots\ldots\text{Eq. 9.19}$$

$$= 4.3964/65.6$$

$$= 0.067018$$

Therefore our linear regression equation is:

$$y = 0.2798 + 0.067018(x - 4.2)$$
$$y = -0.001677 + 0.06702x \quad \ldots\ldots\ldots\ldots\text{Eq. 9.20}$$

9.4.3 Interpolation of the unknowns

To facilitate interpolation of the unknowns, the last cited equation may be rearranged as:

$$x = \frac{y + 0.001677}{0.06702} \quad \ldots\ldots\ldots\ldots\ldots\ldots\text{Eq. 9.21}$$

We can then substitute our mean responses ($\bar{y}$) for unknowns A, B and C from Table 9.4, with the following results:

Unknown	mean ($\bar{y}$)	x (mg) by Eq. 9.21	x (mg) by graph
A	0.046	0.711	0.70
B	0.280	4.202	4.1
C	0.501	7.50	7.4

where it will be seen that the algebraic and graphical values are very close. However, the virtue of the algebraic method is its objectivity, in that it does not depend on fitting the best line by eye and then making a visual judgement of the interception points. Furthermore, the algebraic method permits the estimation of confidence limits, as now follows.

9.4.4 95% confidence limits (approximate)

In an assay such as the one under consideration where the dose–response curve is quite steep and the points show relatively little scatter, it is acceptable to calculate the 95% confidence limits by an approximate formula. However, if these conditions are in doubt or if the exact confidence limits are desired, then the more complex formula of the next section should be used.

The formula for the *approximate* 95% confidence limits (CL) is:

$$CL = \bar{x}_s + m \pm \frac{t \cdot s}{b} \sqrt{\left(\frac{1}{N_s} + \frac{1}{N_u}\right) + \left(\frac{m^2}{\Sigma x_s^2 - (\Sigma x_s)^2/N_s}\right)} \quad .\text{Eq. } 9.22$$

In our example, $\bar{x}_s = 4.2$ from the footnote of Table 9.4.

$$m = \frac{\bar{y}_u - \bar{y}_s}{b} = \frac{\bar{y}_u - 0.2798}{0.067018}$$

and we then have to substitute the mean response ($\bar{y}_u$) of each unknown in turn. t is the Student t-statistic for $P = 0.05$ and the number of d.f. associated with the standard deviation (5 in this case). From the t-table, $t = 2.571$

s = standard deviation (from the residual variance of Table 9.5)
   $= 0.01223$
b = slope $= 0.067018$
N_s = total number of observations on the standard $= 10$
N_u = total number of observations on an unknown $= 2$
$\Sigma x_s^2 - (\Sigma x_s)^2/N_s = 65.6$ from Table 9.4.

Before starting to substitute into the equation, we can do some preliminary calculations:

$$\frac{t \cdot s}{b} = 2.571 \times 0.01223/0.067018$$

$$= 0.4692$$

$$\frac{1}{N_s} + \frac{1}{N_u} = \frac{1}{10} + \frac{1}{2} = 0.6$$

The values of m, $(\bar{x}_s + m)$ and m^2 for each unknown are:

Unknown	$\bar{y}_u$	m	$(\bar{x}_s + m)$	m^2
A	0.046	−3.4886	0.7114	12.1704
B	0.280	0.002984	4.2030	0.0000089
C	0.501	3.30061	7.5006	10.89400

It is now straightforward to calculate the confidence limits by Eq. 9.22 and, for example, with unknown A, the calculation is:

$$CL = 0.7114 \pm 0.4692\sqrt{(0.6 + 12.1704/65.6)}$$
$$= 0.30, \; 1.13$$

Similarly for the other unknowns, we have:

for *B:* CL = 3.84, 4.57

for C: CL = 7.09, 7.91

There are two comments to be made about these results.

(1) When interpolating unknown responses on to the standard dose–response line, the narrowest confidence limits are obtained in the middle of the line and they then flare out towards the end. Thus, for maximum precision it is best to choose, if possible, doses of the unknowns which will give responses close to $\bar{y}_s$. This is because in such cases the term $(\bar{y}_u - \bar{y}_s)/b$ becomes zero, or close to it, and the m^2 term under the square root in Eq. 9.22 disappears.

(2) Other factors which lead to a narrowing of the confidence limits are:

- high value of b
- small value of s
- small value of t
- large values of N_s and N_u

In practice, however, b and s are unlikely to be alterable at the will of the investigator and the main pathway towards narrower CL is to increase the number of observations on standard and unknown. This will also lower the value of t by increasing the number of d.f.

9.4.5 95% confidence limits (exact)

Calculation of the exact 95% confidence limits requires introduction of a term g which is defined by:

$$g = \left(\frac{t \cdot s}{b} \right)^2 \cdot \left(\frac{1}{\Sigma x_s^2 - (\Sigma x_s)^2/N_s} \right) \quad \ldots\ldots \text{Eq. 9.23}$$

Substituting the previously obtained values, we get:

$$g = \frac{(0.4692)^2}{65.6}$$

$$= 0.00336$$

In practice, unless g gets as large as about 0.05, we can ignore it, in the sense that the formula for exact confidence limits will yield virtually the same answer as the approximate formula for $g \leqslant 0.05$. However, let us go ahead nevertheless and calculate the exact limits and compare them with the approximate limits already obtained.

In fact, what we want is not g, but $1 - g$. Here $(1 - g) = 1 - 0.00336 = 0.99664$.

We then substitute it into the formula for the exact 95% confidence limits:

$$CL = \bar{x}_s + \frac{m}{(1-g)} \pm \frac{t \cdot s}{b(1-g)} \sqrt{(1-g) \left(\frac{1}{N_s} + \frac{1}{N_u} \right) + \left(\frac{m^2}{\Sigma x_s^2 - (\Sigma x_s)^2 / N_s} \right)}$$

$$\ldots\ldots\ldots\text{Eq. 9.24}$$

Note that this formula reduces to Eq. 9.22 for the approximate limits, if we put $g = 0$.

We can make use of values already calculated for the approximate formula and, for example, the exact limits for unknown A are:

$$4.2 + \frac{-3.4886}{0.99664} \pm \frac{0.4692}{0.99664} \sqrt{0.99664\ (0.6) + \frac{12.1704}{65.6}}$$

which works out at 0.28 and 1.12. These may be compared with the approximate limits of 0.30 and 1.13.

As a matter of routine, it is not much more difficult to calculate the exact limits each time and, especially if you are going to use a computer programme, this would seem to be the reasonable thing to do.

9.4.6 Additional considerations

In setting up standard curves, many investigators have a fairly casual approach to choosing the number of doses or concentrations, their spacing and the number of replicates at each dose. In the Biuret test, for example, why did we set up doses at 1, 2, 4, 6 and 8 mg of standard protein? If from experience we know that the response is linear between 0 and 8 mg, why do we not simply set up replicates of 0 and 8 mg? Or at most, say 0, 2 and 8? Probably with the Biuret test it does not matter very much; but we should ask what happens in a system where each observation is expensive and difficult to make? Suppose we can only afford to make 12 observations to set up the standard curve. Would we do best to have 12 different concentrations with 1 observation at each? Or 6 concentrations with 2 replicates at each, or 3 concentrations with 4 replicates at each, and so on? The answer is that if we are *sure* of the shape of the dose–response relationship, then it is best (from a precision standpoint) to have fewer different concentrations with more replicate observations on each, than to have the converse. On the other hand, if we have doubts about always hitting the linear region of the dose–response relationship (e.g. in a system that shows day-to-day variation), then it would be better to have a sufficient series of doses to span the desired concentration range, but still to have at least 2 observations at each.

Another important point in actually doing the assay is the avoidance of bias. So it is necessary to have some system of randomizing the sequence of setting up and processing the various doses of standard and unknown. For example, in a colorimetric assay where the colour may change with time, false results might emerge from an assay in which the various unknowns and standard were not intermingled on the time scale.

9.5 Radioimmuno Assays

Because of their widespread use in the biomedical sciences and their unusual dose–response curve, radioimmuno assays (RIA) deserve some mention here. As with other assay systems, the numerical evaluation of RIA data can be approached at different levels of sophistication, ranging from the simple graphical methods to complicated computer programs. The latter may be complex in their construction but simple to apply at the user level. It is not proposed here to discuss RIA in detail, but simply to highlight some of the special features. Brief mention, within the context of dealing with count data, has already been made in Section 6.3.2.

The basic procedure in RIA is to expose constant amounts of the appropriate specific antibody to constant amounts of radiolabelled antigen in the presence of variable amounts of standard unlabelled antigen. After equilibration, the antibody-bound radioactivity in each mixture is then physically separated from the unbound, or 'free' radioactivity and both may be counted. A standard curve is plotted of the dose of unlabelled antigen against one of a choice of count variables notably:

(1) B/F, the ratio of bound to free radioactivity;

(2) B/B_0, the ratio of radioactivity bound in the presence of a particular dose of standard antigen to that bound with zero dose of standard antigen;

(3) B/T, the ratio of bound radioactivity to the total initially added.

Curves of different shapes are obtained according to which plot is adopted, but each allows the interpolation of count ratios from test samples run in parallel with the standard and hence yields estimates of antigen concentration in the test samples. As with other purely graphical methods it is not possible to attach confidence limits to the estimates obtained.

The algebraic evaluation of RIA data has reached a high level of sophistication and the reader is referred to specialized reference to get further details (e.g. Rodbard, 1974). The essential first step is to linearize the relationship between antigen concentration and the count ratio so as to permit fitting the best straight line by the method of least squares. This requires a double transformation: antigen concentration is expressed in logarithms to the base 10 and the count ratios (B/T) are transformed to the quantities known as logits. B/T can take on any value between 0 and 1.0 and the logit of B/T is defined by:

$$logit \left(\frac{B}{T} \right) = ln \left(\frac{B/T}{1 - B/T} \right) \quad \ldots \ldots \ldots Eq.\ 9.25$$

where ln refers to natural logarithms (to the base e). The logit scale runs from about -3.0 through 0 to about $+3.0$, corresponding to B/T ratios from about 0.05 to 0.95. For example, a B/T ratio of 0.05 has a

$$\text{logit} = \ln \left(\frac{0.05}{1 - 0.05} \right)$$

$$= \ln 0.05/0.95$$

$$= \ln 0.05263$$

$$= -2.944$$

A large proportion of RIA assay systems can be linearized by plotting the logit of B/T, or of B/B_0, against $\log_{10}$ concentration of antigen—what is known as the log-logit transformation.

A feature of the log-logit system is that although the dose–response relationship is linear, it is not 'homoscedastic', i.e. the standard deviation is not uniform down the line, being highest at the ends and smallest in the middle. A consequence of this is that when fitting the linear regression line, a process of 'weighting' the points must be used. Subsequent calculations are best done by a computer program and are therefore not given here.

10

Parallel-Line and Slope-Ratio Assays

All animals are equal, but some animals are more equal than others.

George Orwell (Eric Blair), 1903–1950, Animal Farm

10.1 Introduction to Biological Assay

10.1.1 Ingredients and designs

The purpose of a biological assay, or bioassay, is to measure the activity of a biologically-active substance by the effect that it produces on living matter. Such assays have 4 essential ingredients: (a) the sample being assayed, the *unknown*; (b) another sample of the same substance whose potency is known and which therefore acts as a *standard* or *reference*; (c) a supply of living plants, animals, microorganisms, cells or tissues which act as 'targets' for the standard and the unknown; and (d) the *response*, the observable or measurable alteration in the biological material produced by the different dilutions, or doses, of standard and unknown. Some examples of bioassay ingredients are given in Table 10.1, where the term *stimulus* is used for the biologically active substance together with its mode of administration.

From a statistical standpoint, most biological assays follow one of three main designs:

(1) The *standard-line interpolation* assay, which has the same layout as the Biuret protein assay of Section 9.4 and will not be discussed here

(2) The *parallel-line* assay, in which a straight-line dose–response graph is produced by dilutions of the standard, and a parallel, straight line is produced by dilutions of the unknown. This type of assay comes in various forms, e.g. 4-point, 6-point and quantal-response assays, all of which are described below

Table 10.1 A few examples of stimuli, test organisms and responses in biological assays

Substance	Stimulus — Mode of administration	Test organism or tissue	Response
Penicillin	Wells cut in an inoculated agar plate	*Bacillus subtilis*	Zone of inhibition of growth
Indolylacetic acid	Agar blocks	The growing tips of oat seedlings	Bending
Tetanus toxin	Intramuscular injection	Mouse	Death or paralysis
Histamine	Bathing the organ in a controlled-temperature-bath	Strips of guinea pig ileum	Muscle contraction recorded on a kymograph
DDT	Aerosol	House flies	Death
Serum complement	Mixing with target cells in a test tube	Antibody-coated erythrocytes	Haemolysis
Lipopolysaccharide (endotoxin)	Intravenous injection	Rabbits	Elevation in body temperature

(3) The *slope-ratio* assay, in which the straight-line dose–response graphs of standard and unknown, instead of being parallel, meet at the point of zero concentration of the active substance. An example such an assay is given in Section 10.5.

Bioassays have the virtue of measuring directly the biological activity of the substance in question. Often, however, they can be replaced by chemical or immunological procedures which are usually much cheaper and quicker to do. Insulin, for example, can be assayed very conveniently and rapidly by radioimmunoassay (RIA). However, RIA does not distinguish between insulin itself and pro-insulin, the biologically inactive form of the hormone. Thus, a laboratory doing insulin research would probably want to have both RIA and bioassay procedures in its repertoire. A similar example is provided by tetanus toxin and toxoid. Both react with antibody in various observable ways which provide assay responses, but only the toxin causes muscular paralysis.

10.1.2 Biological standards

A major difficulty in biological assays is the inherent variability of biological material, including its tendency to mutate, and the difficulty in getting consistent and reproducible results. In those assays that have microorganisms as the targets, it is possible to achieve some measure of standardization by specifying use of a particular strain, e.g. *Escherichia coli* NCTC 8878 for the assay of vitamin B12 (Hewitt, 1977). However, this specification is not enough, because even with a standard strain the amount of growth is not determined solely by the amount of B12 supplied, but is also influenced by the sum-total of all the other environmental factors in the culture. There has therefore arisen a *philosophy of biological assay*, the essence of which is that standardization is best achieved by taking a batch of target organisms and comparing their responses to the *unknown* sample with those elicited by a *standard* or *reference* substance administered in parallel. Provided that the standard and unknown are allocated in an unbiased manner, the biological variability is evenly distributed between them and the potency of the unknown can be expressed as a multiple, of fraction, of that of the standard. *A biological standard is not, therefore, a test organism and its response*, but is *a sample of a biologically active substance suitably stored to preserve its activity*. For example, the International Reference Preparation of Streptomycin is a batch of ampoules, each containing 175 mg of streptomycin sulphate. The international unit of Streptomycin is defined arbitrarily as 0.001282 mg of this powder. A difference between a biological standard and a physical standard, such as that for the metre, or the kilogram, is that some of the standard is inevitably used up whenever an assay is done. Therefore an international biological standard is not taken for everyday work, and experimentalists have to prepare or acquire their own house standard for routine use. These local standards can initially be calibrated against the international standard so that the potency is known in international units.

10.2 Four-Point Parallel-Line Assay

10.2.1 Preliminaries

The simplest form of parallel-line assay is the 4-point design in which the response given by a high and low dose of standard are compared with those given by a high and low dose of the unknown. Two parallel dose–response lines are thus produced, and by their horizontal displacement allow the potency of the unknown to be determined.

Let us take as our example the assay of penicillin by its growth-inhibiting effect on *Bacillus subtilis* NCTC 3610. The bacteria are uniformly dispersed in nutrient agar gel in petri dishes, and the penicillin at different dilutions is added in constant volumes to circular wells cut in the agar. During incubation, the antibiotic diffuses from each well into the agar and produces a circular zone of bacterial growth inhibition whose diameter is measured next day. Technical details may be found in Wardlaw (1982).

Before actually setting up a 4-point assay, it is necessary to explore the dose–response characteristics of the system over a broad range to determine the region in which a straight-line relationship exists. Fig. 10.1a shows that the graph of zone diameter against penicillin concentration is hyperbolic, while Fig. 10.1b shows that this curve can be linearized over the range from 0.1 to 10 units per ml by plotting the $\log_{10}$ of the concentration instead of its arithmetic value. This straight-line relationship then allows penicillin solutions of unknown potency to be assayed by the standard curve interpolation method of Section 9.4. The only difference is that a logarithmic scale is used on the x-axis, and the algebraic evaluation would be done with the $\log_{10}$ values of the concentrations.

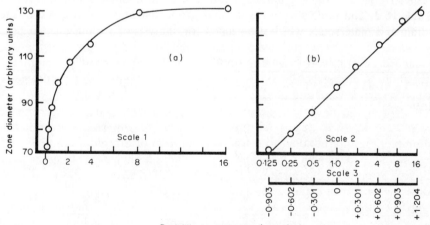

Fig. 10.1   Penicillin dose/response curves. In (a) the plot of antibiotic concentration on an arithmetic scale against the diameter of the zone of growth-inhibition gives an obvious curve; in (b) linearization is achieved by expressing the concentration on a logarithmic scale. This can be done in the two alternative ways illustrated by scales 2 and 3.

Suppose now that a particular solution of penicillin has had its potency determined approximately by the standard curve interpolation method and we now want a more accurate value. This is where the 4-point parallel-line assay is useful.

The design basis of the 4-point assay is that only 2 graph points, provided they are reasonably spaced, are needed to plot the position of a straight line. Therefore to assay a solution of penicillin of unknown concentration against a solution of standard penicillin requires only 2 graph points for each preparation. Therefore each preparation is tested undiluted (the 'high dose') and at, say, a 1/4 dilution (the 'low dose'). This provides 4 graph points which can be joined up so as to give one dose–response line for the standard and another for the unknown. The 2 lines should be parallel. However, the lines will not coincide, unless, by chance, the unknown happens to contain exactly the same concentration of penicillin as the standard. Otherwise, the lines will be horizontally displaced from each other. The relative potency of the unknown in relation to the standard can be determined from the extent and direction of this horizontal displacement.

Let us now work through an actual example in which the high (5.0 units/ml) and low (1.25 units/ml) doses of standard, and high and low doses of unknown, were each tested in quadruplicate. The low dose of unknown was a 1/4 dilution of the high dose, to keep the assay design symmetrical. A convenient physical layout was to have 4 petri dishes, each with 4 wells, and containing one replicate of each dose of each preparation. The 4 solutions were added to each dish in a predetermined random sequence so that, from a statistical viewpoint, each dish was a randomized block (Section 8.3). The whole assay thus consisted of 4 such blocks. The low and high doses of standard and unknown are designated, respectively, S_L, S_H, U_L and U_H, and the zone diameters are set out in Table 10.2. This table also shows the mean zone diameters, as well as the dose totals and plate totals which are needed for analysis of variance.

Table 10.2   Sample results of a 4-point assay of penicillin

Plate	Zone diameter, arbitrary units (y)[a]				Plate total (Q)
	S_L	S_H	U_L	U_H	
A	83	101	79	95	358
B	85	103	81	96	365
C	82	98	80	94	354
D	87	101	78	98	364
Dose total (T)	337	403	318	383	1441
Dose mean $(\bar{y})$	84.25	100.75	79.5	95.75	

[a]Zone diameters were measured by placing each culture plate on an overhead projector and throwing an image, enlarged about 20 times, on to the laboratory wall. An ordinary ruler was then used to read the diameter of each zone image in millimetres (see Wardlaw, 1982).

10.2.2 Graphical evaluation

This is best done on semi-logarthmic paper, as in Fig. 10.2, where the arithmetic scale is used for the responses, as measured by zone diameter, and the logarithmic scale for antibiotic concentrations. Note that these latter cannot be expressed directly in units per ml, because we know the values only for the standard. Instead, we express concentration in *relative volumetric terms*, e.g. a relative concentration of 1.0 designates the undiluted standard or unknown solutions, while a relative concentration of 0.25 describes the 1/4 dilution of each preparation.

In drawing the two dose–response lines we do not simply connect up the mean low and high dose responses of each preparation. Instead, we draw in the two *best-fitting parallel* lines, taking equal account of all 4 points. This is because we are making the underlying assumption that the true dose–response lines *are* parallel, and any departure therefrom is due solely to random-error effects.

Having inserted the two best-fitting parallel lines, we then estimate the potency of the unknown as follows: interpolate upwards from the 1.0 abscissa value to intersect the unknown line, then across horizontally to the standard line and back down to the *x*-axis, following the arrows, as on Fig. 10.2. The point where the arrow returns to the *x*-axis (around 0.67 in this example) gives us the *relative potency* (R) of the unknown. Relative potency is defined as:

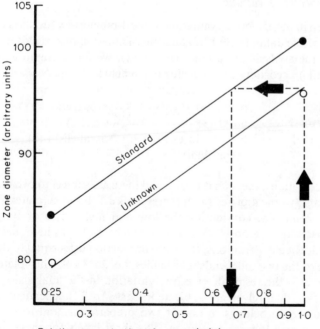

Fig. 10.2 Graphical plot of 4-point assay results from Table 10.2. The horizontal displacement of the two lines is $\log_{10}R$, the logarithm of the relative potency.

$$R = \frac{\text{volume of standard}}{\text{volume of unknown}}$$

which gives the same biological effect. The reason for doing the interpolation with the arrowed lines in this way is to get the arithmetic convenience of 1.0 in the denominator of the R-ratio. In fact, since the lines are parallel, we could draw any horizontal line between them, interpolate from the intersection points down to the x-axis and calculate the ratio of relative concentrations.

If, as here, the unknown solution has an estimated relative potency, from the graph, of 0.67, then its potency in penicillin units per ml is $5.0 \times 0.67 = 3.35$, since the standard penicillin was 5.0 units per ml.

Graphical analysis of the assay data is always worthwhile as a check against gross errors in the algebraic evaluation. For many purposes it gives a sufficiently accurate estimate of potency. Its main drawbacks are that it does not provide an objective assessment of such aspects as significance of block-to-block variation or departures from parallelism; nor does it provide 95% confidence limits for the potency estimate. To explore these aspects we first do an analysis of variance to establish the validity of the assay and then calculate the relative potency and confidence limits by formulae.

10.2.3 Analysis of variance

The first step in the algebraic evaluation of the 4-point assay results is the orderly tabulation of the data (Table 10.2) and the calculation of dose totals and plate totals. For the analysis of variance (A. of V.), we then assign the degrees of freedom (d.f.) according to the following scheme:

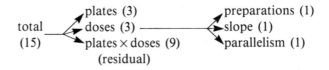

With 16 observations, the total d.f. is 15 and these are subject to two independent classifications: to the 4 plates (with therefore 3 d.f. between plates) and to the 4 doses of penicillin, considering the low and high doses of standard and unknown each to be 'a dose'. Wherever one has results, as here, being subject to 2 independent classifications, there are interaction d.f.s equal to the product of the d.f.s of the two independent variables, i.e. $3 \times 3 = 9$. This 'plates $\times$ doses' variation is also the residual, or error, variation in the analysis.

The 4 doses themselves are subject, in turn, to 2 independent criteria of classification: each belongs to one of two preparations and to either a high concentration or a low concentration. The short labels for these two criteria are 'preparations' with 1 d.f. (since there are 2 preparations) and 'slope' with 1 d.f., since the slope of each dose–response line is fixed by the positions of 2 doses. The *residual* term, which is another interaction term, and is sometimes referred to as *preparation* × *slope*, is more descriptively labelled as 'parallelism' for reasons given below.

Before constructing the A. of V. table we need an *interaction table* for the dose totals. In Table 10.3, the 4 T-values from Table 10.2 are classified according to their low or high dose status and whether they belong to standard or unknown. There is then further row and column totalling to give:

ΣS = sum of all 8 responses given by the standard = 740
ΣU = sum of all 8 responses given by the unknown = 701
ΣL = sum of all 8 responses given by low doses = 655
ΣH = sum of all 8 responses given by high doses = 786

The A. of V. is set out in Table 10.4. The sums of squares (s.s.) in column 3 all have the same correction factor $(\Sigma y)^2/N$ and have as the divisor of their first term the number of observations that went into each group total. Thus, the 'plates' s.s. is derived from Q-values, each of which is a total of 4 observations. Similarly, the divisor in the 'preparations' s.s. is 8 because each sub-total ΣS and ΣU is composed of 8 observations. As with other A. of V., the F-ratios are obtained by dividing each mean square by the residual, or error, mean square. Also, we do not bother to calculate F for all the sources of variation since only some are of interest. Each F-value is associated with a null hypothesis (NH) which is tested by comparing the calculated value of F with the value of F in the F-table for the appropriate number of d.f. Let us deal with each in turn.

Table 10.3 Interaction table for the dose totals

| Preparation | Dose | | Total |
	Low	High	
Standard	337	403	740 (ΣS)
Unknown	318	383	701 (ΣU)
Total	655 (ΣL)	786 (ΣH)	1,441

Between plates. The NH states: 'There is no variation between plates other than that due to random-sampling fluctuations.' The found $F = 2.88$ is less than the tabulated $F = 3.86$ for d.f. = 3,9 and $P = 5\%$. There are thus no grounds for rejecting the NH and we conclude that 'there is no significant between-plates variation'. This is a desirable result.

Slope. The NH states: 'There is no difference between the summed responses to all the low doses and all the high doses, other than that due to random-sampling fluctuations.' The found $F = 458.35$ is very much greater than the tabulated $F = 10.56$ for d.f. = 1,9 and $P = 1\%$. There are therefore compelling reasons for rejecting the NH and we conclude that 'the assay has a highly significant slope'. This is an essential and desirable result for, without it, we would have to *declare the assay invalid* and no further calculations would be worth doing (i.e. this NH must be *rejected* at least at the $P = 5\%$ level in order to proceed further with the calculations).

Table 10.4   Analysis of variance of 4-point assay results

Source of variation	Degress of freedom	Sums of squares	V	F
Total	15	$\Sigma y^2 - CF = 1208.94$ ①	—	—
Between plates	3	$\Sigma Q^2/4 - CF = 20.19$ ②	6.73 ⑧	2.88 (NS)
Between doses	3	$\Sigma T^2/4 - CF = 1167.69$ ③	—	—
Slope	1	$\dfrac{(\Sigma L)^2 + (\Sigma H)^2}{8}$ $- CF = 1072.56$ ④	1072.56 ⑨	458.35**
Preparations	1	$\dfrac{(\Sigma S)^2 + (\Sigma U)^2}{8}$ $- CF = 95.06$ ⑤	95.06 ⑩	40.62**
Parallelism	1	③ − ④ − ⑤ $= 0.07$ ⑥	0.07 ⑪	0.03[a] (NS)
Residual	9	① − ② − ③ $= 21.06$ ⑦	2.34 ⑫	—

CF = correction factor = $(\Sigma y)^2/N$ = 129,780.0625.

[a]As in previous instances of the F-ratio being less than 1.0 (e.g. Table 9.5), this must be inverted for significance testing: $1/0.03 = 33.33$ with 9 and 1 d.f. This is still NS since the tabulated F-values are 240.5 and 6022 for $P = 5\%$ and 1%, respectively.

Preparations. The NH states: 'There is no difference between the summed responses of the two preparations, the standard and the unknown, other than that due to random-sampling fluctuations.' The found $F = 40.62$ is much greater than the tabulated $F = 5.12$ for d.f. = 1,9 and $P = 5\%$, and also greater than the 10.56 tabulated value of F for $P = 1\%$. We can therefore reject this NH and conclude: 'There *is* a significant difference in the potency of the two preparations.' The outcome of this test does not affect the *validity* of the assay but does give us advanced notice that the 95% confidence limits which we shall finally calculate for the potency estimate of the unknown will not overlap with the relative potency value of 1.0 which is assigned to standard.

Parallelism. The NH states: 'There is no deviation from parallelism of the two dose–response lines other than that due to random-sampling fluctuations.' Here the found $F = 0.03$ (which is taken as the reciprocal value 33.3 for significance testing) is much less than the tabulated values given in the footnote of Table 10.4. There are thus no grounds for rejecting the NH and we may conclude that 'the two dose–response lines do *not* deviate significantly from parallelism'. This result is highly desirable because unless the parallelism criterion is met, as it is here, the assay would have to be rejected as *invalid*. The essence of a parallel-line assay is that the lines *should* be parallel or, at least, not deviate significantly (at the $P = 5\%$ level) from parallelism.

This completes the A. of V. and the only quantity that we have to carry forward from the table is the value of standard deviation. This is the square root of quantity ⑫, 2.34, the residual mean square, and equals 1.530 with 9 d.f.

To help in the interpretation of the A. of V. of the 4-point assay, a set of 6 fictitious assays with various graph patterns of dose–response is matched with the corresponding A. of V. (Fig. 10.3). Only assays 2 and 4 are valid, the rest being invalidated by either lack of parallelism (1), lack of slope (3 and 5) or lack of both (6). This diagram is very similar to Fig. 8.2 from a previous chapter.

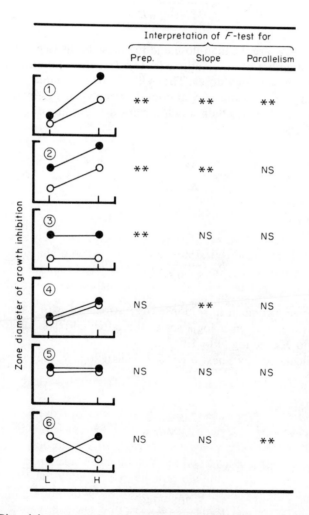

Fig. 10.3 Pictorial representation of main patterns of results of 4-point assays and the corresponding outputs from the analysis of variance ($\bullet$ = standard; $\circ$ = unknown; L = low dose; H-high dose; NS = not significant; ** = highly significant)

10.2.4 Relative potency and confidence limits

If the assay surmounts the hurdle of A. of V. by meeting the criteria of having a significant slope and no significant deviation from parallelism, one may then calculate the relative potency of the unknown with respect to the standard, and the 95% confidence limits.

The first step is to calculate the mean slope (b) of the two dose–response lines from:

$$b = \frac{\Sigma H - \Sigma L}{\frac{1}{2} N \cdot \log_{10} D} \quad \dots\dots\dots\dots\dots\dots\dots \text{Eq. 10.1}$$

where ΣH and ΣL are the high-dose and low-dose totals (see Table 10.3) and N is the total number of observations ($= 16$). D is the dilution factor 4 which separates the low and high doses. The $\frac{1}{2}$ in the denominator arises from the fact that each *pair* of L and H responses contributes *one* estimate of Δy, the numerator in the slope formula which for any straight line of y against x can be expressed as:

$$b = \frac{\Delta y}{\Delta x} \quad \dots\dots\dots\dots\dots\dots\dots\dots \text{Eq. 10.2}$$

In this example

$$b = \frac{786 - 655}{8 \times 0.6021} = 27.1965$$

b is the average increment in response (Δy) associated with a unit increment in the abscissa (Δx). On a $\log_{10}$ scale, the latter corresponds to a 10-fold increase in dose since $\log_{10} 10 = 1.0$.

Having calculated b, we next use it to calculate $\log_{10} R$, the logarithm of the relative potency:

$$\log_{10} R = \frac{\Sigma U - \Sigma S}{\frac{1}{2} N \cdot b} \quad \dots\dots\dots\dots\dots\dots\dots \text{Eq. 10.3}$$

Substituting our figures, we get:

$$\log_{10} R = \frac{701 - 740}{8 \times 27.1965}$$

$$= -0.1793$$

Taking antilogs, we get $R = 0.66$, which is very close to the value $R = 0.67$ which was estimated from the graph.

Since the potency of the standard was 5 units per ml, the estimated potency of the unknown is:

$$5 \times 0.66 = 3.30 \text{ units per ml}$$

In order to calculate the confidence limits, we first need $S_{\log R}$, the standard error of $\log_{10} R$, given by:

$$S_{\log R} = \frac{2s}{b} \sqrt{\frac{1}{N} \left[1 + \left(\frac{\log_{10} R}{\log_{10} D} \right)^2 \right]} \quad \ldots\ldots \text{Eq. 10.4}$$

Note that this is an *approximate* expression and that there is an *exact* formula which may be found in, for example, Finney (1971).

Substituting our numerical values yields:

$$S_{\log R} = \frac{2 \times 1.530}{27.1965} \sqrt{\frac{1}{16} \left[1 + \left(\frac{(-0.1793)}{(\ 0.6021)} \right)^2 \right]}$$
$$= 0.02935$$

Note that a calculation such as this can easily be done on a calculator without writing down any of the intermediate results, so the formula is not nearly as tedious as it might seem at first sight.

It is highly desirable to have $S_{\log R}$ as small as possible because this will give the highest precision (narrowest confidence limits). Looking at the formula it can be seen that $S_{\log R}$ is made small by:

(1) low standard deviation (s);
(2) high slope (b);
(3) large number of observations (N); and
(4) $\log_{10} R = 0$ (i.e. $R = 1$, which means that the unknown is the same potency as the standard. This causes the term $(\log_{10} R/\log_{10} D)^2$ to disappear. Having obtained $S_{\log R}$, the 95% confidence limits (CL) are given by:

$$95\% \text{ CL} = \text{antilog } [\log_{10} R \pm t \cdot S_{\log R}] \quad \ldots\ldots\ldots\ldots\ldots \text{Eq. 10.5}$$

where we make the following substitutions:

$$\log_{10} R = -0.1793$$
$\quad\quad\quad t = 2.262$, the tabulated value of the t-statistic for the number of d.f. associated with s, the standard deviation (in our case d.f. $= 9$) and $P = 5\%$
$$S_{\log R} = 0.02935$$

The lower 95% CL is given by

$$LL = \text{antilog}\ (-0.1793 - 2.262 \times 0.02935)$$
$$= \text{antilog}\ (-0.2457)$$
$$= 0.57$$
$$= 0.57 \times 5 = 2.84 \text{ units per ml}$$

The upper 95% CL is given by

$$UL = \text{antilog}\ (-0.1793 + 2.262 \times 0.02935)$$
$$= \text{antilog}\ (-0.1130)$$
$$= 0.77$$
$$= 0.77 \times 5 = 3.85 \text{ units per ml}$$

Our final result is thus:

Estimate of potency = 3.30 u/ml.

Lower 95% confidence limit = 2.84 u/ml.

Upper 95% confidence limit = 3.85 u/ml.

Note that the confidence limits are distributed unsymmetrically around R, i.e.

$$(R) - (LL) = 3.30 - 2.84 = 0.46$$
$$(UL) - (R) = 3.85 - 3.30 = 0.55$$

A useful spot-check for arithmetic errors in calculating the confidence limits is to see that the ratios

$$\frac{LL}{R} = \frac{R}{UL} \text{ are the same.}$$

With our figures we have:

$$\frac{2.84}{3.30} = 0.86$$

$$\frac{3.30}{3.85} = 0.86$$

which agree to within rounding-off errors.

The above calculation may seem to be protracted but this is only because space has been taken to explain the individual steps. When such assays are being done routinely, one would use a computer program which would give a print-out of all the above calculations in a very short time and which would be free from any possibilities of arithmetic errors.

To recapitulate, the important design elements of a 4-point parallel-line assay are to:

(1) Choose doses in the linear part of the dose—response curve.

(2) Use the same dilution interval, e.g. 4-fold for both standard and unknown.

(3) Set up the same number of replications of each dose of standard and unknown.

(4) Try (e.g. on the basis of a preliminary assay) to arrange for the high dose

of unknown to be as close as possible in estimated potency to the high dose of standard.

(5) Avoid bias by having a complete assay inside each block (plate) and by adding the solutions in an unbiasing manner.

10.2.5 Combining several estimates

If we have several independent assays on a sample it is normal to combine the estimates of relative potency and confidence limits so as to arrive at a composite estimate, taking account of all the results. This should also give narrower confidence limits.

This is *not* done by taking the simple arithmetic average of all the R-values. Instead we use a process of *weighting* which gives extra emphasis to the individual assays of greatest precision (small $S_{\log R}$) and, conversely, downgrades the importance of assays with high values of $S_{\log R}$.

Table 10.5   Combining the results of 5 independent assays

Assay no.	$S_{\log R}$	W	$\log_{10} R$ (R)	$W \log_{10} R$
1	0.0294	1156.92	-0.1793 (0.66)	-207.43
2	0.0764	171.32	-0.1909 (0.64)	-32.705
3	0.066	229.57	-0.1286 (0.74)	-29.522
4	0.0392	650.77	-0.2575 (0.55)	-167.573
5	0.049	416.49	-0.2366 (0.58)	-98.542
Totals	—	2625.07	—	-535.772

Table 10.5 sets out the results of 5 independent assays on an unknown penicillin done by different students in the same class. It will be seen that the values of $\log_{10} R$ range from -0.1286 (corresponding to $R = 0.74$) to -0.2575 (corresponding to $R = 0.55$), and that $S_{\log R}$ ranges from 0.0294 to 0.0764.

For each value of $S_{\log R}$ we calculate a *weight* (W) defined as:

$$W = \frac{1}{(S_{\log R})^2} \quad \dots\dots\dots\dots\dots\dots \text{Eq. 10.6}$$

This has been done in column 3 of the table and shows that assay no. 1 is rated nearly 7 times as 'good' as assay no. 2 by this process of weighting. We also calculate the sum of the weights (ΣW).

Each weight is multiplied by the corresponding value of $\log_{10} R$ to give $W \cdot \log_{10} R$ in the 5th column of the table. We then calculate the sum of the $W \cdot \log_{10} R$-values.

The composite estimate $(\bar{R})$ of relative potency in the 5 assays is given by:

$$\bar{R} = \text{antilog} \left[\frac{\Sigma W \cdot \log_{10} R}{\Sigma W} \right] \quad \dots\dots\dots\dots\text{Eq. 10.7}$$

This works out as

$$\bar{R} = \text{antilog} \left[\frac{-535.772}{2625.07} \right]$$

$$= \text{antilog} \ (-0.204098)$$
$$= 0.6250 \text{ in relative terms}$$
$$= 0.6250 \times 5 = 3.126 \text{ units per ml}$$

The effect of this procedure is to 'swing' the value of $\bar{R}$ closer to the individual estimates of R that have the smallest $S_{\log R}$ associated with them.

To obtain the new confidence limits we must first obtain a composite value for $S_{\log R}$ based on the 5 assays. This quantity is defined by

$$\overline{S_{\log R}} = \sqrt{\frac{1}{\Sigma w}} \quad \dots\dots\dots\dots\dots\dots\dots\text{Eq. 10.8}$$

$$= \sqrt{\frac{1}{2625.07}}$$

$$= 0.01952$$

Note that this is considerably lower than even the lowest of the individual $S_{\log R}$-values.

The new confidence limits are given by

$$CL = \text{antilog} \ (\overline{\log_{10} R} \pm t \cdot \overline{S_{\log R}}) \dots\dots\dots\dots\text{Eq. 10.9}$$

where we take the value of t for the sum of the d.f. of the individual assays. Since each assay had 9 d.f. associated with error, this gives 45 d.f. in total. From tables, the value of t for 45 d.f. is about 2.01. We therefore get:

$$CL = \text{antilog} \ (-0.2041 \pm 2.01 \times 0.01952)$$
$$= 0.57 \text{ and } 0.68$$

for the lower and upper limits, respectively.

These limits are decidedly narrower than those of any of the individual assays and it has therefore proved advantageous to produce the composite estimates.

10.2.6 Standard versus several unknowns

The 4-point design of standard versus a single unknown can readily be extended to permit the assay of several unknowns (U^1, U^2, U^3, etc.) simultaneously against the same standard. However, since the standard will then be subjected to multiple comparisons, it is desirable to have extra data on the standard responses. The rule is that with k unknowns, it is desirable that there should be $\sqrt{k}$ times as many observations on standard as on each individual unknown. Thus, if the assay is of standard versus 4 unknowns, there should be twice as many observations on standard as on each unknown. At the same time, it is desirable to preserve the symmetry of the assay by having the same number of observations at each dose level. Probably the best compromise, to avoid awkward numbers in the A. of V., is to set up the standard in duplicate if there are between 3 and 6 unknowns, and in triplicate if there are 7 or more unknowns.

A statistician should be consulted for assistance with assay design and the changes needed in the various formulae.

10.3 Six-Point Parallel-Line Assay

The 4-point parallel-line assay is fine if you are sure of having all of the responses in the linear region of the log dose–response relationship. With the penicillin assay on culture plates this is easy to arrange. However, in other biological assay systems, particularly those involving animals, there may be a great deal of uncontrollable day-to-day variation in the responsiveness of the test subjects. Here a set of doses, which may be right in the centre of the linear region on one day, may not be so placed on another day, and may give responses which extend into the non-linear parts of the graph. This means that if an assay is inadvertently set up with doses in these regions, the 4-point design will not detect it. Fig. 10.4 illustrates what happens when 4-point and 6-point assays are set up at doses near to the lower end of the dose–response curve where there is curvature. The kinking, which is obvious visually from the graphs, can be confirmed by the analysis of variance (A. of V.) (Section 10.3.3) of the 6-point assay which has a new 'kink-detection' term not present in the 4-point design. Naturally, it is undesirable for *any* parallel-line assay to be set up in curved regions of the dose–response relationship, but at least the 6-point design lets one know if this has happened.

10.3.1 A specific example

As indicated in Fig. 10.4, the 6-point design consists of a low, a middle and a high dose each of standard and unknown. In the interest of symmetry, the dose interval between low and middle, and between middle and high doses, should be the same. For the same reason of symmetry, we should take the same number of observations in each of the 6 dose groups. As with the 4-point design, our discussion here is restricted to test systems where the response variable is

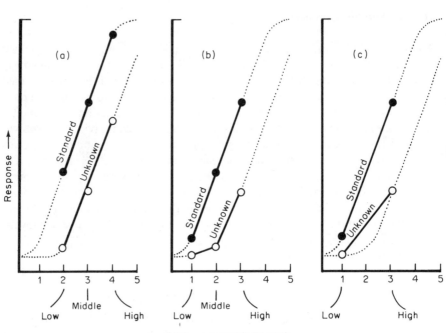

Fig. 10.4 Illustrating the advantage of the 6-point assay design for detecting curvature if, accidentally, the doses are set up in a non-linear region of the dose–response curve. a, 6-point assay set up in the linear region of the dose–response curves; b, 6-point assay set up with the low dose of the unknown in the non-linear zone: analysis of variance would detect 'quadratic curvature'; c, as b except middle doses omitted, thus giving a 4-point assay. This does not distinguish between a deviation from parallelism and an underlying bend in the line

continuous. Later in the chapter we shall consider a 6-point assay where the response variable is discontinuous and where a probit transformation is needed. Meanwhile, let us take as our working example a 6-point parallel-line assay on penicillin, with 4 replicates at each dose. The conduct of the assay will be very similar to the 4-point design discussed above. Let us further suppose that the $6 \times 4 = 24$ wells were filled with the 6 doses by a scheme of full randomization, i.e. treating the 6 petri dishes, with 4 wells on each, as one large agar surface. We shall use a dose interval of 3 between the low and middle, and the middle and high doses.

10.3.2 Tabulation of results and graphical evaluation

Table 10.6 sets out the results of the 6-point assay of standard penicillin at 5.0 units per ml against a 'known–unknown' of true potency 3.0 units per ml. By 'known–unknown' is meant an unknown which has been prepared by an accurate dilution of the standard and which therefore provides a useful check on the quality of the assay. For example, knowledge of the true potency of the

Table 10.6   Results of 6-point assay of standard penicillin versus a 'known–unknown'. Zone diameters are measured in arbitrary units

	Standard (5.0 u/ml)			'Unknown' (3.0 u/ml)		
	Low S_L	Middle S_M	High S_H	Low U_L	Middle U_M	High U_H
	77	92	110	73	84	100
	75	94	102	71	85	104
	76	90	106	73	86	97
	73	91	106	67	89	100
Dose Totals	301 T_1	367 T_2	424 T_3	284 T_4	344 T_5	401 T_6
Means	75.25 $\bar{y}1$	91.75 $\bar{y}2$	106 $\bar{y}3$	71 $\bar{y}4$	86 $\bar{y}5$	100.25 $\bar{y}6$

The dose interval between successive dilutions is 3.

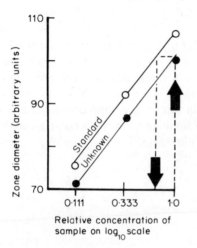

Fig. 10.5   Plot of dose–response lines of 6-point, parallel-line assay (data of Table 10.6)

unknown can be withheld from the person doing the assay until the assay has been finished and all the calculations are complete. The estimated relative potency, R, can then be compared with the true relative potency. At the base of Table 10.6 are the dose totals (T) ready for A. of V. and the dose means for plotting the dose-response graph. In this graph (Fig. 10.5) the points have been plotted on semi-logarithmic paper with the abscissa labelled in the same way as the 4-point assay (Fig. 10.2). The best parallel straight lines have then been fitted by eye and it will be noted that they do not pass exactly through all of the plotted points. This is to be expected. By extrapolating upwards from the abscissa value 1.0 to the U line, then horizontally to S and back down to the dose axis, we get a value of $R = 0.64$ as our graphical estimate of relative potency. This corresponds to $5 \times 0.64 = 3.20$ units per ml.

10.3.3 Analysis of variance

The graphs in Fig. 10.5 suggest that the assay is unlikely to be invalidated by either lack of parallelism or presence of kinking. Nevertheless we should seek confirmation by performing the A. of V., which will also give us a value of s, the standard deviation.

Since we have used a fully randomized design for assigning the doses to the petri dishes, there is no 'block' component in the analysis of variance. Instead, our total of 23 d.f. is divided as follows:

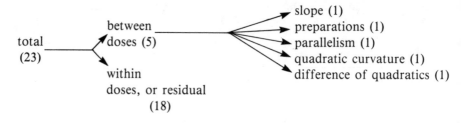

The first branch point is straightforward in that the 24 observations with 23 d.f., are split into 5 d.f. between doses and 18 d.f. within doses. The doses themselves are not just 6 random doses, but are doses that bear a special, classifiable relationship to each other. The associated 5 d.f. are, in fact, assignable to the 5 distinct analytic components listed at the second branch point above. The first 3 are familiar from the 4-point assay, while the last 2 reflect the possibility for 'kink-detection' in the 6-point design. Thus, dose–response lines like:

or

would have a significant F-ratio for 'quadratic curvature', whereas lines like this:

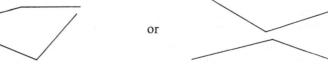

or

would have a significant F-ratio for 'difference of quadratics'.

The key to the A. of V. of the 6-point assay is to make use of coefficients of orthogonal contrasts, as previously employed in Section 8.5.

For a symmetrical 6-point assay with n observations per dose, the scheme to be used is that in Table 10.7. No explanation for the theoretical basis of orthogonal contrasts will be attempted. They are simply offered as empirical tools. But we can check that the coefficients are *contrasts* and that they are *orthogonal* by showing that they meet the following two criteria:

Table 10.7   Coefficients of orthogonal contrasts for a symmetrical 6-point parallel-line assay

Dose	S_L	S_M	S_H	U_L	U_M	U_H	Divisor for sum of squares	Symbol for the row sum
Response totals	T_1	T_2	T_3	T_4	T_5	T_6		
Slope	-1	0	1	-1	0	1	$4n$	$L1$
Preparations	-1	-1	-1	1	1	1	$6n$	Lp
Parallelism	1	0	-1	-1	0	1	$4n$	$L1'$
Quadratic curvature	1	-2	1	1	-2	1	$12n$	$L2$
Difference of quadratics	-1	2	-1	1	-2	1	$12n$	$L2'$

$n =$ number of observations in each T value.

(1) each set of coefficients (each horizontal row) shall add to zero, e.g.

$$L1 = -1 + 0 + 1 - 1 + 0 + 1 = 0$$
$$L2' = -1 + 2 - 1 + 1 - 2 + 1 = 0$$

(2) each pair of contrasts shall be orthogonal, i.e. the products of corresponding coefficients for any two contrasts shall add to zero. For example, if we take $L1$ and $L2'$ as a pair we get:

$$(-1)(-1) + (0)(2) + (1)(-1) + (-1)(1) + (0)(-2) + (1)(1)$$
$$= 1 - 1 - 1 + 1$$
$$= 0$$

The first application of the contrasts is in the A. of V. Thus the sum of squares (s.s.) for 'slope' is $(L1)^2/4n$, where $L1$ is calculated from the T-values in Table 10.6 and the coefficients in Table 10.7 as:

$$L1 = -301 + 0 + 424 - 284 + 0 + 401 = 240$$

where each entry, like -301, is obtained by multiplying the T-value ($T_1 = 301$) by the corresponding coefficient (-1). Likewise, the other L-values are:

$$Lp = -301 - 367 - 424 + 284 + 344 + 401 = -63$$
$$L1' = 301 + 0 - 424 - 284 + 0 + 401 = -6$$
$$L2 = 301 - 2 \times 367 + 424 + 284 - 2 \times 344 + 401 = -12$$
$$L2' = -301 + 2 \times 367 - 424 + 284 - 2 \times 344 + 401 = 6$$

These L-values are then squared, divided by the divisor $4n$, $6n$ or $12n$ (where n is the number of measurements per dose total), as listed in Table 10.7, and entered into the A. of V. table (Table 10.8) in the s.s. column. It will be noted that in the A. of V. we have not bothered to calculate the 'correction factor' $(\Sigma y)^2/N$. This is because it is not actually needed unless we wish to confirm that the subdivisions of the 'doses' s.s. do add up to the 'doses' s.s. as shown in the footnote to the table.

Table 10.8   Analysis of variance of 6-point assay

Source	Degrees of freedom	Sums of squares[a]		V	F
Total	23	Σy^2	= 191,327.0 ①	—	—
Doses	5	$\Sigma T^2/4$	= 191,214.75 ②	—	—
Slope	1	$(L1)^2/16 =$	3,600.0	3,600	577
Preparations	1	$(Lp)^2/24 =$	165.375	165.375	26.52
Parallelism	1	$(L1')^2/16 =$	2.25	2.25	0.36[b]
Quadratic curvature	1	$(L2)^2/48 \ =$	3.0	3.0	0.48[b]
Difference of quadratics	1	$(L2')^2/48 =$	0.75	0.75	0.12[b]
Residual	18	① − ② =	112.25	6.236	—

[a]Note that the 'correction factor' $(\Sigma y)^2/N$ is not calculated since it is not required, except for the purpose of checking that the sums of squares calculated with the L-values adds up to the sums of squares for 'doses'. For this purpose we need to subtract $(\Sigma y)^2/N$ from 191,214.75.
$\Sigma y = 2121$; $(\Sigma y)^2/N = 187,443.375$
$\Sigma T^2/4 - (\Sigma y)^2/N = 3771.375$
which is the sum of the slope, parallelism, etc. terms that are the subdivisions of the 'doses' sums of squares.
[b]Note that reciprocals of these values must be used for significance testing and that none is significant.

The column of F-ratios in Table 10.8 shows that the assay has a highly significant slope and that there is a highly significant difference in the potency of standard and unknown. Moreover, there is no suggestion of the assay being invalidated on the possible grounds of lack of parallelism or presence of quadratic curvature or of a difference of quadratics. We can therefore proceed to work out the mean slope, relative potency and confidence limits. The standard deviation, taken as the square root of the residual mean square, is 2.4972, with 18 d.f.

10.3.4   Slope, relative potency and confidence limits

Implicit in the use of coefficients of orthogonal contrast is a 'metameric' transformation of the x-axis of the dose–response graph to logarithms to the base D, where D is the dilution interval. In this assay $D = 3$. As a result of this metameric transformation, the low, middle and high doses can be represented on a dimensionless scale as -1, 0 and $+1$, which is why the transformation is done. We should not allow ourselves to be baffled by attempts to visualize logs to the base D, because at any time we can convert results into logs to the base 10 by multiplying or dividing by $\log_{10}D$. In practice, it is convenient to carry the calculations of slope, relative potency and confidence limits in terms of logs to the base D and then convert the results right at the end. A benefit of using this procedure is that the equations themselves are simpler.

Using the value of the $L1$ orthogonal contrast, b', the slope expressed in $\log_D$ is given by:

$$b' = \frac{L1}{4n} \quad \dots \dots \dots \dots \dots \dots \dots \dots \dots \dots \dots \dots \dots \text{Eq. 10.10}$$

$$= \frac{240}{4 \times 4}$$

$$= 15$$

Numerically, this means that a unit change on the metameric scale, such as going from the low dose to the middle dose, is accompanied on average by an increment of 15 in zone diameter. This can easily be confirmed by subtracting any two adjacent means on the bottom line of Table 10.6.

If we want the slope on a $\log_{10}$ scale, we can divide b' by $\log_{10}3 = 0.4771$ to get 31.44 as the average increment in response for a unit change on the $\log_{10}$ scale, i.e. a 10-fold increase in dose. But meanwhile let us remain with the $\log_D$ scale where $b' = 15$.

The relative potency (M) on the $\log_D$ scale is given by:

$$M = \frac{Lp}{3nb'} \quad \dots \dots \dots \dots \dots \dots \dots \dots \dots \dots \dots \dots \text{Eq. 10.11}$$

$$= \frac{-63}{3 \times 4 \times 15}$$

$$= -0.35$$

Here to get $\log_{10}R$, we *multiply* by $\log_{10}D$, i.e.

$$\log_{10}R = -0.35 \times 0.4771$$
$$= -0.1670$$

Taking antilogs gives R, the relative potency, as 0.68, or 3.40 units per ml.

Let us now go straight to calculating the exact 95% confidence limits, which is simplified by working in logs to the base D.

First, we need the Fieller's term (g), defined as:

$$g = \frac{1}{4n} \left(\frac{t \cdot s}{b'} \right)^2 \quad \dots \dots \dots \dots \dots \dots \dots \dots \dots \text{Eq. 10.12}$$

where t is the tabulated value of the t-statistic for $P = 0.05$ and the degrees of freedom associated with s.

$$g = \frac{1}{4 \times 4} \left(\frac{2.10 \times 2.4972}{15} \right)^2$$

$$= 0.0076$$

Therefore $(1 - g) = 0.9924$.

The exact lower and upper 95% confidence limits expressed in logs to the base D are designated M_L and M_U and are given by:

$$M_L, M_U = \frac{1}{(1-g)}\left[M \pm \frac{t \cdot s}{b'} \sqrt{(1-g)\frac{2}{3n} + \frac{M^2}{4n}} \right] \dots\dots\dots\dots\text{Eq. 10.13}$$

$$= \frac{1}{0.9924}\left[-0.35 \pm \frac{2.10 \times 2.4972}{15} \sqrt{\frac{0.9924 \times 2}{3 \times 4} + \frac{(-0.35)^2}{4 \times 4}} \right]$$

$$= -0.4992 \text{ and } -0.2061$$

Multiplying by $\log_{10}D = 0.4771$ gives $\log_{10}LL$ and $\log_{10}UL = -0.2382$ and -0.0983. Taking antilogs to base 10 gives the lower and upper confidence limits as 0.58 and 0.80 on the relative potency scale. To convert to units per ml, multiply by the 5 u/ml in the undiluted standard, to get values of 2.9 and 4.0 for the confidence limits of the 'unknown'. Note that these enclose the true value of 3.0.

10.4 Quantal Response Assay

In the 4-point and 6-point designs just considered, the response variable is *continuous* in the sense that the zone diameters of growth inhibition can be measured to as many decimal places as desired. Many other types of bioassay also have continuously variable responses, e.g. physiological responses such as growth, blood pressure, body temperature or the concentration of a metabolite. In other bioassays, however, the respose is *quantal*, i.e. all-or-nothing, and in Chapter 5 we presented dose–response lines and methods for estimating ED_{50}.

The position becomes considerably more complex when we have to deal with 2, or more, parallel dose–response lines and wish to determine the potencies of unknown samples relative to a standard. As with other assays, we can do a rough graphical plot — in this case, the probit of the percentage of responders versus the logarithm of the dose. The assay can be either 4-point or 6-point, and one can estimate the log of the relative potency by the horizontal displacement of the dose–response lines of the unknowns from that of the standard.

Where we get into difficulties is in taking the calculations further, i.e. fitting the best lines by weighted linear regression, doing an analysis of variance and working out relative potencies and confidence limits. For the purposes of this book, the calculations are just too long and tedious to be presented here and, as a matter of practicality, one would never attempt them with a pocket calculator. The reader who needs these procedures and who has a microcomputer, or access to a mainframe computer is therefore recommended to use such facilities. Larsson, Harboe and Aalen (1981) have produced computer programmes for this purpose, written in BASIC, FORTRAN and PASCAL. A sample printout of results of a pertussis-vaccine potency assay is given in

Table 10.9 Computer printout of a quantal response assay of standard versus 3 unknowns, each preparation being tested at 3 doses in groups of 19 or 20 animals per dose

```
****************
** DOSE-RESPONSE **
**  EVALUATION  **
**     OF       **
**  PERTUSSIS   **
**   VACCINE    **
****************
```

DATE OF IMMUNISATION: FEBR. 26TH, 1981
ITERATION LIMIT: .1%
PROTECTIVE UNITS OF STANDARD: 3

NUMBER OF ITERATIONS: 4
COMMON SLOPE: .796

STANDARD

DOSE	ANIMALS	RESPONSE	CALC. RESP.
.05	20	20	19.22
.01	20	15	13.7
.002	20	2	4.24

STANDARD   ED-50 .00546   SE. ED-50 .00138

TEST 1

DOSE	ANIMALS	RESPONSE	CALC. RESP.
.05	19	19	18.61
.01	20	14	15.52
.002	20	7	6.01

TEST 1   ED-50 .00386   SE. ED-50 .00104
PROTECT. UNITS 4.24   SE. P.U. 1.52
95% CONFIDENCE-LIMITS 2.11–8.54

TEST 2

DOSE	ANIMALS	RESPONSE	CALC. RESP.
.05	19	18	18.74
.01	19	15	15.6
.002	20	9	7.17

TEST 2   ED-50 .00315   SE. ED-50 .000898
PROTECT. UNITS 5.19   SE. P.U. 1.89
95% CONFIDENCE-LIMITS 2.55–10.6

TEST 3

DOSE	ANIMALS	RESPONSE	CALC. RESP.
.05	20	18	17.57
.01	20	8	9.08
.002	19	2	1.55

TEST 3   ED-50 .0116   SE. ED-50 .00284
PROTECT. UNITS 1.42   SE. P.U. .498
95% CONFIDENCE-LIMITS .712–2.82

CHI-SQUARE: 7.4   LIMIT: 14.06   D.F. 7

Reproduced by permission of the National Institute of Public Health, Postuttak, Oslo 1, Norway.

Table 10.9. It shows a test of standard versus 3 unknowns, each being tested at 3 dose levels in groups of 19 or 20 animals. The computer printout gives the ED_{50} of each preparation and the protective potency of each unknown ('test') in units per ml, the standard vaccine having a value of 3.0. The chi-square value at the bottom right is an index of assay validity analogous to an analysis of variance.

10.5 Slope-Ratio Assay

In the slope-ratio type of assay, the standard and the unknown both show straight-line dose–response graphs which, instead of being parallel, meet at the focal point of zero dose. Each straight line requires 2 points to define its position and, independently, the two straight lines so defined have to pass through the point of zero dose, i.e. the 'blank' value. Therefore the simplest type of slope-ratio assay is the 5-point design with replicate observations at each point. Let us now work through an example. For other designs, see Finney (1971).

10.5.1 Protocol and results

Table 10.10 presents the results of a 5-point assay of the concentration of nicotinic acid in an extract of dried apricots. The assay was done by putting two different amounts (in nanograms) of standard nicotinic acid into sets of triplicate test tubes, and two different amounts (in microlitres) of apricot extract into other triplicated sets of tubes. In addition there were triplicates of 'blank' tubes with diluent only. To all tubes was then added a constant amount of nicotinic-acid-free culture medium and an inoculum of washed cells of *Lactobacillus plantarum* which requires nicotinic acid as a specific growth factor. The tubes were incubated for 72 hours and the extent of bacterial growth measured turbidimetrically. Technical details may be found in Wood and Norris

Table 10.10 Protocol and results of a 5-point, slope-ratio, microbiological assay of the concentration of nicotinic acid in an extract of dried apricots

| | Table of optical density (E_{540}) readings | | | | |
| | Standard[a] | | | Unknown[b] | |
	Blank B	50 ng S_L	100 ng S_H	200 μl U_L	400 μl U_H
	0.03	0.30	0.59	0.27	0.43
	0.01	0.34	0.61	0.24	0.39
	0.02	0.29	0.63	0.21	0.45
Total	0.06	0.93	1.83	0.72	1.27
	(T_1)	(T_2)	(T_3)	(T_4)	(T_5)

[a]Nicotinic acid.
[b]An extract of 5 g dried apricots in 1 litre.

(1982). Note that some small degree of turbidity is expected in the blank tubes because of difficulty in keeping them absolutely free of all traces of nicotinic acid from the culture medium and the bacteria.

10.5.2 Graphical evaluation

Unlike parallel-line assays where the dose scale is logarithmic, slope-ratio assays have a simple arithmetic scale of dose, going from zero to the highest dose used. For this reason the low dose of each preparation was made one-half of the high dose, so as to achieve even spacing on the arithmetic scale. Fig. 10.6 gives a plot of the assay results. Since the standard and unknown were dispensed in different units (nanograms and microlitres, respectively), 2 different abscissa scales are used. However, they are brought into comparability by having the high dose values coincident, as shown. Underneath these two is a dimensionless *metameric* scale which allows us to link the nanogram and microlitre scales. It has a range from 0 to 1.0.

Slope is defined as the increase in response given by a unit increase in dose. If we take for this purpose an increment of 1.0 on the metameric scale, then

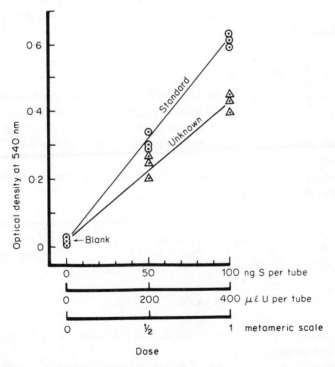

Fig. 10.6   Dose–response lines of a 5-point slope-ratio assay of the concentration of nicotinic acid in a sample of dried apricots. Note the 3 different ways of labelling the abscissa

from the graph the response increment of standard is approximately $0.615 - 0.02 = 0.595$, and for the unknown is $0.425 - 0.02 = 0.405$.

The slope-ratio (R) is:

$$R = \frac{b_u}{b_s} \qquad\qquad\qquad\qquad\qquad\text{Eq. 10.14}$$

where b_u and b_s are the slopes of unknown and standard, respectively. Substituting the graph values gives:

$$R = 0.405/0.595$$
$$= 0.68$$

The calculation is completed by expressing the concentration of nicotinic acid in the unknown by the following argument: at the top dose of unknown the $400\,\mu l$ must have contained 100×0.68 ng of nicotinic acid, since the high dose of standard was 100 ng and the relative potency of the unknown is 0.68. Therefore the concentration of nicotinic acid in the unknown is $68\,ng/400\,\mu l$, which is equivalent to 170 ng/ml or $170\,\mu g/l$. As indicated in the footnote to Table 10.10, this solution was obtained by extracting 5 g dried apricots and making the volume up to 1 litre. Therefore the fruit is estimated to contain $170/5 = 34\,\mu g$ nicotinic acid per gram.

10.5.3 Analysis of variance

As with other assay designs, the analysis of variance (A. of V.) serves the twin functions of providing tests of validity, together with an estimate of the error standard deviation, s. The best way to do the analysis is to use coefficients of orthogonal contrast to help in the calculation of the subdivision of the 'doses' sum of squares.

For a symmetrical 5-point assay with n observations per dose, the degrees of freedom (d.f.) diagram is:

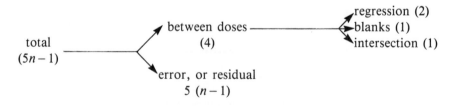

The meaning of the new terms 'blanks' and 'intersection' will emerge as we perform the analysis.

As with other A. of V. the first step is to produce an orderly tabulation of the data, with treatment totals (T-values) calculated (Table 10.10). We then consult Table 10.11 which contains both regression and contrast coefficients

Table 10.11   Coefficients of regression and orthogonal contrasts for the
5-point slope-ratio assay

Symbol[a] for row sum	Coefficient for dose-total					
	B T_1	S_L T_2	S_H T_3	U_L T_4	U_H T_5	Divisor[b]
L_S	-15	1	17	-6	3	$17.5n$
L_U	-15	-6	3	1	17	$17.5n$
L_B	2	-2	1	-2	1	$14n$
L_I	0	2	-1	-2	1	$10n$

[b] n = number of observations at each dose.
[a] L_S and L_U are regression contrasts while L_B and I_I are orthogonal contrasts.

and is based on Finney (1971, p. 204). The first two entries, L_s and L_u, are for the calculations of slopes of the 2 dose–response lines and are not used in the A. of V. The other 2 quantities, L_B and L_I, are, however, used in the A. of V.

L_B is the 'blanks' contrast whose value is calculated by multiplying the T-values from Table 10.10 by the contrast coefficients given in the body of Table 10.11, thus:

$$L_B = (2 \times 0.06) - (2 \times 0.93) + (1 \times 1.83) - (2 \times 0.72) + (1 \times 1.27)$$
$$= -0.08$$

Similarly the 'intersection' contrast (L_I) is obtained as:

$$L_I = (0 \times 0.06) + (2 \times 0.93) - (1 \times 1.83) - (2 \times 0.72) + (1 \times 1.27)$$
$$= -0.14$$

The rest of the A. of V. is straightforward and follows the general scheme of Table 10.12. The numerical output from the analysis is given in Table 10.13,

Table 10.12   Analysis of variance table for 5-point slope-ratio assay with n observations
at each dose

Source of variation	Degreees of freedom	Sums of squares[a]	V	F
Total	$5n-1$	$\Sigma y^2 - CF$ .... ①	—	—
Doses	4	$\Sigma T^2/n - CF$ ..... ②	—	—
'Blanks'	1	$(L_B)^2/14n$ .... ③	③	③/⑦
'Intersection'	1	$(L_I)^2/10n$ .... ④	④	④/⑦
Regression	2	② − ③ − ④ .. ⑤	⑤/2	⑤/⑦
Residual	$5(n-1)$	① − ② .... ⑥	⑥/5 $(n-1)$ = ⑦	

Standard $(s) = \sqrt{⑦}$

[a] $CF = (\Sigma y)^2/5n$.

Table 10.13   Analysis of variance of the nicotinic acid assay by inserting the results from Table 10.10 into the scheme of Table 10.12

Source of variation	Degrees of freedom	Sum of squares	V	F
Total	14	0.579893	—	—
Doses	4	0.573827	—	—
'Blanks'	1	0.000152	0.000152	0.25[a] (NS)
'Intersection'	1	0.000653	0.000653	1.08 (NS)
Regression	2	0.573022	0.286511	472 (**)
Residual	10	0.006066	0.0006066	—

$s = 0.02463$.

[a]As explained previously (e.g. Table 8.5), this ratio must be inverted for significance testing. When this is done, $1/0.25 = 4.0$ with 10 and 1 d.f. Since the tabulated F for d.f. 10, 1 is 241.9 and 6056, the result is still not significant.

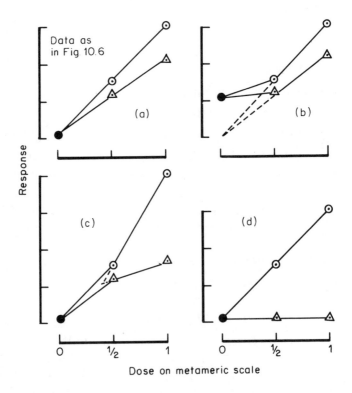

Fig. 10.7   Valid (a) and invalid (b and c) slope–ratio assay dose–response relationships. In (b), the 'blanks' term of the analysis of variance is significant, while in (c), the 'intersection' term is significant. (d) is not invalid, but because the 'unknown' line is horizontal, the relative potency is zero

For retention of validity, it is necessary for the 'blanks' and 'intersection' terms to be non-significant. Fig. 10.7 illustrates various forms of invalidity. In Fig. 10.7b, where the 'blanks' term is significant, the actual dose–response lines both have a kink in the same direction, which means that the response is not linear down to zero dose. In Fig. 10.7c, where the 'intersection' term is significant, the kinkings of the two lines are in opposite directions, so that the apparent 'focal point' of the lines is horizontally separated from the zero-dose point. Fig. 10.7d presents a valid assay, but where the unknown has zero potency, as shown by its zero slope. Note that the regression term in the analysis only becomes non-significant if *both* dose–response lines have non-significant slopes.

At the bottom of the A. of V. table we get the standard deviation $s = 0.02463$ to be carried forward to the next section.

10.5.4 Slope and relative potency

The simplest way to calculate the slopes of the 2 dose–response lines is to make use of the regression contrasts in Table 10.11. The slope of standard (b_s) is given by:

$$b_s = L_s/17.5n \dots\dots\text{Eq. 10.15}$$

Substituting figures from Table 10.10 gives:

$$b_s = \frac{(-15 \times 0.06) + (1 \times 0.93) + (17 \times 1.83) + (-6 \times 0.72) + (3 \times 1.27)}{17.5 \times 3}$$

$$= 0.5834$$

which is close to the value 0.595 obtained graphically. Similarly, the slope of the unknown (b_u) is given by:

$$b_u = L_u/17.5n \dots\dots\text{Eq. 10.16}$$

$$b_u = \frac{(-15 \times 0.06) + (-6 \times 0.93) + (3 \times 1.83) + (1 \times 0.72) + (17 \times 1.27)}{17.5 \times 3}$$

$$b_u = 0.4061$$

which again is very close to the value (0.405) obtained graphically. From Eq. 10.14, the relative potency (R) is:

$$R = 0.4061/0.5834$$
$$= 0.6961$$

which is within 3% of the graphical value 0.68.

Those who pursue the mathematics of the slope-ratio assay in Finney's (1971) book will see that we are dealing here with *multiple linear regression*, in that the 2 dose–response lines are represented by a *single* equation. This follows from the fact that in terms of the underlying model the two lines are not independent of each other since they both should pass through the same blank value.

10.5.5 Confidence limits

As with the parallel-line assays, the confidence limits can be estimated either approximately or exactly, the latter involving a more complex formula.

The approximate formula for the 95% confidence limits for a 5-point slope-ratio assay is:

$$LL, UL = R \pm \frac{t}{b_s} \sqrt{\frac{8s^2}{35n} (8 - 9R + 8R^2)} \quad \dots\dots\dots\dots\dots Eq.\ 10.17$$

where $t = 2.228$ in this example, i.e. the tabulated value for $P = 0.05$ and d.f. $= 10$, the degrees of freedom associated with s.

This gives:

$$LL, UL = 0.6961 \pm \frac{2.228}{0.5834} \sqrt{\frac{8 \times 0.02463^2}{35 \times 3} (8 - 9 \times 0.6961 + 8 \times 0.6961^2)}$$

$$= 0.63 \text{ and } 0.76$$

However, the use of Eq. 10.18 depends on Fieller's term g being negligible, i.e. less than about 0.05.

To check this, we should therefore calculate g from:

$$g = \frac{64}{35n} \left(\frac{t \cdot s}{b_s}\right)^2 \quad \dots\dots\dots\dots\dots\dots\dots\dots\dots\dots\dots Eq.\ 10.18$$

Substituting gives:

$$g = \frac{64}{35 \times 3} \left(\frac{2.228 \times 0.02463}{0.5834}\right)^4$$

$$= 0.0054$$

which, indeed, is negligible and the use of Eq. 10.18 is therefore justifiable.

If, however, g had not been negligible, we should have calculated the confidence limits by the exact formula which is only slightly more complicated than Eq. 10.18 and should, therefore, be used routinely. This formula for exact 95% limits is:

$$UL, LL = \frac{1}{1-g} \left[R - \frac{9g}{16} \pm \frac{t}{b_s} \sqrt{\frac{8s^2}{35n} \left(8 - 9R + 8R^2 - \frac{175g}{32}\right)} \right]$$

$$\dots\dots\dots Eq.\ 10.19$$

which gives exact 95% confidence limits of LL, $UL = 0.635$ and 0.759 which are very close to the values obtained previously.

10.5.6 Other applications

Classically, the slope-ratio design tends to be illustrated by its application to the microbiological assay of the B-vitamins. In principle, however, many other systems would be adaptable to this design. For example, in the Biuret test for protein, which was discussed in Section 9.4 as a standard-curve interpolation assay, the slope-ratio design would be applicable. Instead of having just one dose of unknown, you would have two; and instead of defining the standard line with several points, two would be enough. These two sets of points with each of the standard and the unknown would, together with the blank, give a 5-point slope-ratio design.

Other possible applications are the assay of haemolytic antibodies by a gel diffusion test in agar containing erythrocytes and subsequently flooded with complement (Hiramoto et al., 1970). The diffusion–precipitin test for antigens and antibodies in agar gels, due to Mancini et al. (1965), would also seem to be adaptable to the slope-ratio approach.

References

Ashby, R. E. and Rhodes-Roberts, M. E. (1976). 'The analysis of variance to examine the variations between samples of marine bacterial populations'. *J. appl. Bacteriol.*, **41**, 439–451.

Banic, S. (1975). 'Prevention of rabies by vitamin C.' *Nature Lond.*, **258**, 153–154.

Bliss, C. I. (1967). *Statistics in Biology*, Vol. 1. McGraw-Hill, New York.

British Pharmacopoeia, Vol. 2 (1980). A186–190. Her Majesty's Stationery Office, London.

British Standards Institution (1962). *Specification for Graduated Pipettes* B.S.700: 1962, with subsequent amendments. British Standards Institution, London.

Campbell, R. C. (1974). *Statistics for Biologists*, 2nd edn. Cambridge University Press, Cambridge.

Cinader, B., Dubiski, S. and Wardlaw, A. C. (1966). 'Genetics of MuB1 and of a complement defect in inbred strains of mice.' *Genet. Res. Camb.*, **7**, 32–43.

Cochran, W. G. (1950). 'Estimation of bacterial densities by means of the most probable number''. *Biometrics*, **6**, 105–116.

Colquhoun, D. (1971). *Lectures on Biostatics*. Clarendon Press, Oxford.

Cooke, D., Craven, A. H. and Clarke, G. M. (1982). *Basic Statistical Computing*. Edward Arnold, London.

Demerec, M. (1948). 'Origin of bacterial resistance to antibiotics.' *J. Bacteriol.*, **56**, 63–74.

Fannin, T. E., Marcus, M. D., Anderson, D. A. and Bergman, H. L. (1981). 'Use of a fractional factorial design to evaluate interactions of environmental factors affecting biodegradation rates.' *App. environm. Microbiol.*, **42**, 936–943.

Fegan, R. C. and Brosche, S. L. (1979). *BASIC Computer Programming for Statistics*. Prentice-Hall Inc., Englewood Cliffs, New Jersey.

Finney, D. J. (1962). *Probit Analysis*, 2nd edn. Cambridge University Press, Cambridge.

Finney, D. J. (1971). *Statistical Methods in Biological Assay*, 2nd edn. Griffin, London.

Finney, D. J., Latscha, R., Bennett, B. M. and Hsu, P. (1963). *Tables for Testing Significance in a 2×2 Table*. Cambridge University Press, Cambridge.

Fisher, R. A. (1970). *Statistical Methods for Research Workers*, 14th edn. Oliver and Boyd, Edinburgh, p. 1.

Fisher, R. A. and Yates, F. (1963). *Statistical Tables for Biological, Agricultural and Medical Research*, 6th edn. Oliver and Boyd, Edinburgh.

Furman, B. L., Wardlaw, A. C. and Stevenson, L. Q. (1981). '*Bordetella* pertussis-induced hyperinsulinaemia without marked hypoglycaemia: a paradox explained.' *Brit. J. exp. Path.*, **62**, 504–511.

Herbert, D., Phipps, P. J. and Strange, R. E. (1971). 'Chemical analysis of microbial cells', pp. 209–344 in: J. R. Norris and D. W. Ribbons (Eds.), *Methods in Microbiology, Vol. 5B*. Academic Press, London.

Hewitt, W. (1977). *Microbiological Assay*. Academic Press, New York.

Hiramoto, R. N., McGhee, J. R., Hurst, D. C. and Hamlin, N. M. (1970). 'A simple method for quantitation of hemolytic antibodies.' *Immunochemistry*, **7**, 992–996.

Hirano, S. S., Nordheim, E. V., Arny, D. C. and Upper, C. D. (1982). 'Lognormal distribution of epiphytic bacterial populations on leaf surfaces'. *App. Env. Microbiol.*, **44**, 695–700.

Jones, J. G. (1973). 'Use of nonparametric tests for the analysis of data obtained from preliminary surveys: a review. *J. Appl. Bacteriol.*, **36**, 197–210.

Kolk-Vegter, A., Kolk, A. H. J., Krediet, R. T. and Staugaart-Kloosterziel, W. (1973). 'How good is your serial microdilution technique?' *J. Immunol. Methods*, **3**, 375–384.

Kurokawa, M., Ishida, S., Goto, N. and Kuratsuka, K. (1974). 'A new method for biological assay of endotoxin using changes in peripheral leukocyte population in mice as a response.' *Jap. J. Med. Sci. Biol.*, **27**, 173–189.

Larsson, P. G., Harboe, A. and Aalen, O. O. (1981). 'A computer programme for the evaluation of dose response data.' *Natl. Inst. of Public Health* (Oslo) *Annals*, **4**, 19–31.

Lee, J. D. and Lee, T. D. (1982). *Statistics and Numerical Methods in BASIC for Biologists*. Van Nostrand Reinhold Co., Wokingham, Berks.

Liddell, F. D. K. (1978). 'Evaluation of survival in challenge experiments'. *Microbiol. Rev.*, **42**, 237–249.

Lutz, W. (1967). 'Statistical methods as applied to immunological data,' pp. 1163–1206. in: D. M. Weir (Ed.), *Handbook of Experimental Immunology*. Blackwell, Oxford.

Mackay, A. L. (1977). in: M. Ebison (Ed.), *The Harvest of a Quiet Eye*. Institute of Physics, Bristol.

Mancini, G., Carbonara, A. O. and Heremans, J. F. (1965). 'Immunochemical quantitation of antigens by single radial immunodiffusion.' *Immunochemistry*, **2**, 235–254.

Medical Research Council (1951). 'The prevention of whooping cough by vaccination'. *Brit. Med. J.*, (**1**), 1463–1471.

Medical Research Council (1956). 'Vaccination against whooping cough.' *Brit. Med. J.*, (**2**) 454–462.

Medical Research Council (1959). 'Vaccination against whooping cough.' *Brit. Med. J.*, (**1**) 994–1000.

Meynell, G. G. and Meynell, E. (1970). *Theory and Practice in Experimental Bacteriology*, 2nd edn. Cambridge University Press, Cambridge.

Nimmo, I. A. and Atkins, G. L. (1979). 'The statistical analysis of non-normal (real?) data. *Tr. in Biochem. Sci.*, **4**, 236–239.

Norman, R. L. and Kempe, L. L. (1960). 'Electronic computer solutions for the MPN equation used in the determination of bacterial populations.' *J. biochem. microbiol. Technol. Engng.*, **2**, 157–163.

Norris, K. P. and Powell, E. O. (1961). 'Improvements in determining total counts of bacteria.' *J. Roy. Microscop. Soc.*, **80**, 107–119.

Reid, D. B. W. (1968). 'Statistical methods in virology', pp. 104–124 in: A. J. Rhodes and C. E. van Rooyen (Eds.), Textbook of Virology, 5th edn. Williams and Wilkins, Baltimore.

Rodbard, D. (1974). 'Statistical quality control and routine data processing for radioimmunoassay and immunoradiometric assays.' *Clin. Chem.*, **20**, 1255–1270.

Rowe, R., Todd, R. and Waide, J. (1977). 'Microtechnique for most-probable-number analysis.' *App. environm. Microbiol.*, **33**, 675–680.

Russek, E. and Colwell, R. R. (1983). 'Computation of most probable numbers.' *Appl. environm. Microbiol*, **45**, 1646–1650.

Snedecor, G. W. and Cochran, W. G. (1967). *Statistical Methods*, 6th edn. Iowa State University Press, Ames, Iowa.

Sokal, R. R. and Rohlf, F. J. (1969). *Biometry*, W. H. Freeman & Co., San Francisco.

Taylor, J. (1962). 'The estimation of numbers of bacteria by tenfold dilution series.' *J. app. Bacteriol.*, **25**, 54–61.

Taylor, L. R. (1961). 'Aggregation, variance and the mean.' *Nature Lond.*, **189**, 732–735.

Wardlaw, A. C. (1982). 'Four-point parallel-line assay of penicillin, pp. 370–379, in: S. B. Primrose and A. C. Wardlaw (Eds.), *Sourcebook of Experiments for the Teaching of Microbiology*. Academic Press, London and New York.

Welborn, T. A., Rubenstein, A. H., Haslam, R. and Fraser, R. (1966). 'Insulin response to glucose.' *Lancet*, **1**, 280.

Wiener, A. S. (1943). *Blood Groups and Transfusion*, 3rd edn. p. 190. Charles C. Thomas, Illinois, p. 190.

Wood, B. J. B. and Norris, F. W. (1982). 'Microbiological assay of a vitamin-nicotinic acid', pp. 214–223 in: S. B. Primrose and A. C. Wardlaw (Eds.), *Sourcebook of Experiments for the Teaching of Microbiology*. Academic Press, London and New York.

Glossary

Abscissa. By convention, the horizontal axis of a graph drawn to show the relationship between two variables such as x and y.

Algorithm. A process, or set of rules, for performing a calculation.

Anova, Anovar. Short ways of saying *analysis of variance* (A. of V.).

Array. An arrangement of numerical data in some kind of regular sequence or pattern, e.g. in order of magnitude.

Asymptotic. Refers to a curve which continually approaches either the x or y axis but does not actually reach the axis until x or y equals infinity. An example is the normal distribution curve. The axis so approached is the *asymptote*.

Attribute data. Also known as classificatory data; such data arise from counts of objects or persons possessing some feature that permits classification into two or more discrete categories, such as male/female, dead/alive, sterile/contaminated, etc.; see Section 5.1.

Average. In ordinary usage, the arithmetic mean.

Bar chart. Same as histogram; see Chapter 2.

Bias. Some influence which makes a statistical result non-representative.

Central limit theorem. An important concept which justifies the prominent place occupied by the normal distribution in statistical theory. It states that if you take random samples from *any* population (not necessarily a normal distribution), the *means* of the samples will be approximately normally distributed, a tendency which will increase as the sample size is increased. For many distributions, the means of samples of only 10 observations have a distribution very close to the normal.

Coefficient of variation. A measure of variability defined as the standard deviation (s) multiplied by 100 and divided by the mean ($\bar{y}$), i.e. $CV = 100s/\bar{y}$.

Combinations. The number of combinations of n objects taken r at a time is written nCr and is given by:

$$nCr = \frac{n!}{r!\,(n-r)!}$$

e.g. the number of combinations of a, b, c, d taken 2 at a time is $\dfrac{4!}{2!(4-2)!} = 6$,

i.e. ab, ac, ad, bc, bd, cd. Note that ab is the same *combination* as ba but not the same *permuation*.

Contingency. The difference in the cells of a Contingency Table (see Index) between the actual and expected frequencies.

249

Control chart. A graph which shows the results of repeated sampling during a manufacturing or routine testing process. The chart consists of a central horizontal line representing the average, and horizontal lines ('control limits') above and below this at, for example, ± 1 and ± 2 standard deviations. A set of points outside the control limits suggests that some new influence has crept into the procedures.

Crossover assay. A type of bioassay in which the same individual animals can be used repeatedly and where, to avoid bias, each animal in turn can be exposed to the various doses of standard and unknown; an example is the rabbit blood-sugar method for insulin.

Data. Observations or measurements which are taken as fact and used as the basis of calculations or reasoning. Note that *data* is the plural of *datum*, and one should avoid such solecisms as 'The data shows . . . '.

Decile. If the scale of the measured variable, *y*, of the objects in a population is divided into 10 equal parts so as just to enclose the population, each of the equal divisions is a *decile*.

Degrees of freedom (d.f.). A term widely used in statistics and best explained by examples. The argument goes as follows. Suppose you have taken a sample consisting of *n* independent observations and from these you calculate the mean, $\bar{y}$. Then, having calculated and knowing the value of $\bar{y}$, although you still have your *n* observations, the number of them which are *independent* is now only $(n - 1)$. For example, with 2 observations, say 5.0 and 7.0, once you have calculated $\bar{y} = 6.0$, then you are left with only one *independent* observation, either the 5.0 or the 7.0 — because having chosen one of them, the other is automatically fixed, i.e. is no longer independent.

In the formula for standard deviation, it makes sense therefore to have the divisor $(n - 1)$ instead of *n*, because with *two* observations you only have *one* estimate of the degree of scatter. Thus, it would be false to think you had two *independent* estimates of scatter just because you subtracted each observation in turn from the mean.

It can be seen that the central idea in the term *degrees of freedom* is the notion that a sample has a certain inherent amount of information in it, which is dependent on the sample size, and you cannot manufacture new information just by estimating parameters. We can look at it this way, that the *n* bits of data have *n* degrees of freedom before we do any statistical calculations, but as soon as we estimate a parameter such as $\bar{y}$, we 'use up' one of the d.f. initially present.

In the case of a *t*-test on 2 samples with n_1 and n_2 observations, we 'use up' 2 d.f. in applying the formula, since we calculate *two* means, $\bar{y}_1$ and $\bar{y}_2$, in order to do the test. Therefore when we look up the *t*-table, we enter it at $(n_1 + n_2 - 2)$ d.f.

Likewise in fitting a straight line through paired data points, we 'use up' 2 d.f. in so doing, because the process of line-fitting is a matter of extracting the parameters *a* and *b* for the linear equation $y = a + bx$. In a contingency table with *r* rows and *c* columns, the d.f. $= (r - 1)(c - 1)$.

Degrees of freedom are used in the *t*-, *F*- and χ^2-tests, in analysis of variance, and in calculating variance, standard deviation, standard error of the mean and 95% confidence limits. They do not occur in the non-parametric tests such as Mann–Whitney, Wilcoxon, etc. since it is inherent in these tests that we avoid estimating parameters from the sample data.

Deviation. The difference between an observed value and some standard value, or the mean of the sample.

Direct assay. A bioassay procedure which is done in much the same way as the chemist's titration of an acid with a base. The classical example is of the drug digitalis which, when infused into a cat's heart in small increments, but quite quickly, eventually causes the heart beat to stop. Direct assays are not much used in biological work because of the non-instantaneous response of most biological systems to incremental additions of biologically-active substances.

Error of first kind. Referring to a test of significance, such as the *t*-test, if the null hypothesis is rejected when it really ought to have been accepted. However, to be able to detect such an error implies having some information additional to that around which the null hypothesis was framed; see Sections 3.4.2 and 3.4.3.

Error of second kind. The converse of the above, i.e. if the null hypothesis is accepted when it should have been rejected; see Sections 3.4.2 and 3.4.3.

Estimate, estimator. Two terms with related but different meanings: when the observations in a sample are substituted into appropriate equations to calculate, for example, mean and standard deviation, then the values so obtained are *estimates* (of the underlying population parameters μ and σ) and the equations used are *estimators*.

Exponential. A variable such as x is related to another variable such as y in an exponential fashion if the equation describing the relationship contains x as a power of e, the base of natural logarithms. A simple exponential function is $y = e^x$, which may also be written $y = \ln x$ or $y = 2.303 \log_{10} x$.

Gamma function. Algebraic quantities which occur in the equations that describe the shape of the t, F and χ^2 distributions (see Bliss, 1967).

Geometric mean. The appropriate average or mean for *log-normally* distributed data. The geometric mean of n observations y_1, y_2, y_3, . . ., y_n is antilog $((1/n) \ \Sigma \log y)$.

Harmonic mean. Of a set of numbers, y_1, y_2, y_3, etc. the harmonic mean is the reciprocal of the arithmetic mean of the reciprocals of the numbers, i.e.

$$H = N/\Sigma \ (1/y)$$

Apparently not much used in biology.

Moment. Not used in statistics to mean a brief period of time but as a technical term for the deviations of observations from a mean. The 'first moment' is $(y - \bar{y})$, the second moment $(y - \bar{y})^2$ and so on. The second moment provides the squares of the deviations that are used to calculate the standard deviation. The third and fourth moments are used to investigate the extents of skewness and kurtosis in a distribution.

Monte Carlo method. A procedure for solving a mathematical problem with the assistance of some kind of random-sampling operation. An example is provided in Section 3.4.1, where the approximate shape of the t-distribution for 3 d.f. is obtained in this way.

Moran's test. A procedure for calculating most probable numbers in bacterial dilution counts; see Meynell and Meynell (1970, p. 190).

Murphy's Law. Also know in experimental science as Sod's Law; describes the operation of a Guiding Influence in laboratory work, and indeed, in human affairs generally; commonly formulated as: 'if it is possible for something to go wrong, then sooner or later it will, and usually at the least convenient moment'.

Normal deviate. In a normally-distributed population, it is the difference between an observed value and the mean. See also **standardized normal deviate**.

Ordinate. By convention, the vertical axis of a graph drawn to show the relationship between two variables such as x and y.

Orthogonal, Orthogonality. That feature in the design of an experiment which ensures that the different factors which might influence the variable are capable of direct and separate estimation. The opposite of orthogonal is *confounded*.

Parametric. Adjective applied to those statistical tests, e.g. the t-test, which assume knowledge of the shape of the underlying frequency distribution of the *population* from which the samples were taken.

Permutation. A permutation of n different objects taken r at a time is denoted by nPr or $P(n,r)$ and is given by:

$$nPr = \frac{n!}{(n-r)!}$$

e.g. the number of permutations of the letters a, b, c, d, taken 2 at a time, is $4!/2! = 12$. They are ab, ba, ac, ca, ad, da, bc, cd, bd, db, cd, dc. See also **combinations**.

Pie chart. A circular diagram with radial lines which divide up its area into sectors whose areas represent the relative frequency of occurrence of particular objects or events; popular with epidemiologists for displaying, for example, the relative annual incidence of different diseases in a community.

Point estimate. This is the estimated value of a parameter such as the mean, and is distinct from an *interval estimate*, such as the 95% confidence limits, which replace the point with a band of values in which the parameter is estimated to lie.

Quartile. Values of a variable which divide a population into quarters. Similarly, *deciles* divide into tenths.

Range. The difference between the highest and the lowest measurement in a sample.

Standardized normal deviate. Any measurement (y) which is taken from a normally distributed population can be converted into a standardized normal deviate by substituting it into $Z = (y - \bar{y})/s$, where s is the standard deviation. This has the effect of changing the measurement units in which y was originally measured into dimensionless units whose size is equal to 1 standard deviation. This changes the abscissa of the normal distribution curve into a scale with a central value of $Z = 0$ when $y = \bar{y}$.

Tukey test. A test used in conjunction with analysis of variance to decide whether the data taken for analysis should first be transformed in some way (e.g. into logs); see Snedecor and Cochran (1967, p. 331).

Variable, Variate. Any quantity which varies. Some writers attempt to distinguish between the two terms.

Variance. The sum of the squares of the deviations of the observations from the sample mean, divided by the number of degrees of freedom, i.e.

$$V = \frac{\Sigma(y - \bar{y})^2}{N - 1}$$

Note that this is different from the usage in *accountancy*, where variance means the difference between an actual expenditure and the budgeted figure.

Weibull plot. A graph of log % response against log dose, used in virulence titrations and having the capacity to detect heterogeneity of responsiveness in the test subjects. See Meynell and Meynell (1970, p. 195).

Appendix: Statistical Tables and Figures*

*Tables in Appendices A2, A4, A6, A7 and A17 are taken from Tables III, IV, IX and IX2 of Fisher & Yates: Statistical Tables for Biological, Agricultural and Medical Research published by Longman Group Ltd, London (previously published by Oliver and Boyd Ltd, Edinburgh) and by permission of the authors and publishers.

Appendix A1. Random Digits

20 17	42 28	23 17	59 66	38 61	02 10	86 10	51 55	92 52	44 25
74 49	04 49	03 04	10 33	53 70	11 54	48 63	94 60	94 49	57 38
94 70	49 31	38 67	23 42	29 65	40 88	78 71	37 18	48 64	06 57
22 15	78 15	69 84	32 52	32 54	15 12	54 02	01 37	38 37	12 93
93 29	12 18	27 30	30 55	91 87	50 57	58 51	49 36	12 53	96 40
45 04	77 97	36 14	99 45	52 95	69 85	03 83	51 87	85 56	22 37
44 91	99 49	89 39	94 60	48 49	06 77	64 72	59 26	08 51	25 57
16 23	91 02	19 96	47 59	89 65	27 84	30 92	63 37	26 24	23 66
04 50	65 04	65 65	82 42	70 51	55 04	61 47	88 83	99 34	82 37
32 70	17 72	03 61	66 26	24 71	22 77	88 33	17 78	08 92	73 49
03 64	59 07	42 95	81 39	06 41	20 81	92 34	51 90	39 08	21 42
62 49	00 90	67 86	93 48	31 83	19 07	67 68	49 03	27 47	52 03
61 00	95 86	98 36	14 03	48 88	51 07	33 40	06 86	33 76	68 57
89 03	90 49	28 74	21 04	09 96	60 45	22 03	52 80	01 79	33 81
01 72	33 85	52 40	60 07	06 71	89 27	14 29	55 24	85 79	31 96
27 56	49 79	34 34	32 22	60 53	91 17	33 26	44 70	93 14	99 70
49 05	74 48	10 55	35 25	24 28	20 22	35 66	66 34	26 35	91 23
49 74	37 25	97 26	33 94	42 23	01 28	59 58	92 69	03 66	73 82
20 26	22 43	88 08	19 85	08 12	47 65	65 63	56 07	97 85	56 79
48 87	77 96	43 39	76 93	08 79	22 18	54 55	93 75	97 26	90 77
08 72	87 46	75 73	00 11	27 07	05 20	30 85	22 21	04 67	19 13
95 97	98 62	17 27	31 42	64 71	46 22	32 75	19 32	20 99	94 85
37 99	57 31	70 40	46 55	46 12	24 32	36 74	69 20	72 10	95 93
05 79	58 37	85 33	75 18	88 71	23 44	54 28	00 48	96 23	66 45
55 85	63 42	00 79	91 22	29 01	41 39	51 40	36 65	26 11	78 32
67 28	96 25	68 36	24 72	03 85	49 24	05 69	64 86	08 19	91 21
85 86	94 78	32 59	51 82	86 43	73 84	45 60	89 57	06 87	08 15
40 10	60 09	05 88	78 44	63 13	58 25	37 11	18 47	75 62	52 21
94 55	89 48	90 80	77 80	26 89	87 44	23 74	66 20	20 19	26 52
11 63	77 77	23 20	33 62	62 19	29 03	94 15	56 37	14 09	47 16
64 00	26 04	54 55	38 57	94 62	68 40	26 04	24 25	03 61	01 20
50 94	13 23	78 41	60 58	10 60	88 46	30 21	45 98	70 96	36 89
66 98	37 96	44 13	45 05	34 59	75 85	48 97	27 19	17 85	48 51
66 91	42 83	60 77	90 91	60 90	79 62	57 66	72 28	08 70	96 03
33 58	12 18	02 07	19 40	21 29	39 45	90 42	58 84	85 43	95 67
52 49	40 16	72 40	73 05	50 90	02 04	98 24	05 30	27 25	20 88
74 98	93 99	78 30	79 47	96 92	45 58	40 37	89 76	84 41	74 68
50 26	54 30	01 88	69 57	54 45	69 88	23 21	05 69	93 44	05 32
49 46	61 89	33 79	96 84	28 34	19 35	28 73	39 59	56 34	97 07
19 65	13 44	78 39	73 88	62 03	36 00	25 96	86 76	67 90	21 68
64 17	47 67	87 59	81 40	72 61	14 00	28 28	55 86	23 38	16 15
18 43	97 37	68 97	56 56	57 95	01 88	11 89	48 07	42 60	11 92
65 58	60 87	51 09	96 61	15 53	66 81	66 88	44 75	37 01	28 88
79 90	31 00	91 14	85 65	31 75	43 15	45 93	64 78	34 53	88 02
07 23	00 15	59 05	16 09	94 42	20 40	63 76	65 67	34 11	94 10
90 08	14 24	01 51	95 46	30 32	33 19	00 14	19 28	40 51	92 69
53 82	62 02	21 82	34 13	41 03	12 85	65 30	00 97	56 30	15 48
98 17	26 15	04 50	76 25	20 33	54 84	39 31	23 33	59 64	96 27
08 91	12 44	82 40	30 62	45 50	64 54	65 17	89 25	59 44	99 95
37 21	46 77	84 87	67 39	85 54	97 37	33 41	11 74	90 50	29 62

From Lindley, D. V. and Miller, J. C. P. (1953), *Cambridge Elementary Statistical Tables*, p. 12. *Reproduced by permission of Cambridge University Press.*

Appendix A2. The Student t-Statistic

Degrees of freedom	Value of t for probability (%)				
	10	5	2	1	0.1
1	6.314	12.706	31.821	63.657	636.619
2	2.920	4.303	6.965	9.925	31.598
3	2.353	3.182	4.541	5.841	12.924
4	2.132	2.776	3.747	4.604	8.610
5	2.015	2.571	3.365	4.032	6.869
6	1.943	2.447	3.143	3.707	5.959
7	1.895	2.365	2.998	3.499	5.408
8	1.860	2.306	2.896	3.355	5.041
9	1.833	2.262	2.821	3.250	4.781
10	1.812	2.228	2.764	3.169	4.587
11	1.796	2.201	2.718	3.106	4.437
12	1.782	2.179	2.681	3.055	4.318
13	1.771	2.160	2.650	3.012	4.221
14	1.761	2.145	2.624	2.977	4.140
15	1.753	2.131	2.602	2.947	4.073
16	1.746	2.120	2.583	2.921	4.015
17	1.740	2.110	2.567	2.898	3.965
18	1.734	2.101	2.552	2.878	3.922
19	1.729	2.093	2.539	2.861	3.883
20	1.725	2.086	2.528	2.845	3.850
21	1.721	2.080	2.518	2.831	3.819
22	1.717	2.074	2.508	2.819	3.792
23	1.714	2.069	2.500	2.807	3.767
24	1.711	2.064	2.492	2.797	3.745
25	1.708	2.060	2.485	2.787	3.725
26	1.706	2.056	2.479	2.779	3.707
27	1.703	2.052	2.473	2.771	3.690
28	1.701	2.048	2.467	2.763	3.674
29	1.699	2.045	2.462	2.756	3.659
30	1.697	2.042	2.457	2.750	3.646
40	1.684	2.021	2.423	2.704	3.551
60	1.671	2.000	2.390	2.660	3.460
120	1.658	1.980	2.358	2.617	3.373
∞	1.645	1.960	2.326	2.576	3.291

Appendix A3A. The Variance Ratio (F): 5% Significance Level

Note that the F-ratio must always be calculated with the greater mean square (or variance) placed as the numerator of the fraction. This sometimes means that in analysis of variance, the error variance may occasionally have to be inverted from its normal position in the denominator. When this is done, the insertion of degrees of freedom into the rows and columns of the F-table has to be adjusted appropriately.

Degrees of freedom of lesser mean square (denominator)	Value of F for 5% probability and with the greater mean square having degrees of freedom:														
	1	2	3	4	5	6	7	8	9	10	12	15	20	24	30
1	161.4	199.5	215.7	224.6	230.2	234.0	236.8	238.9	240.5	241.9	243.9	245.9	248.0	249.1	250.1
2	18.51	19.00	19.16	19.25	19.30	19.33	19.35	19.37	19.38	19.40	19.41	19.43	19.45	19.45	19.46
3	10.13	9.55	9.28	9.12	9.01	8.94	8.89	8.85	8.81	8.79	8.74	8.70	8.66	8.64	8.62
4	7.71	6.94	6.59	6.39	6.26	6.16	6.09	6.04	6.00	5.96	5.91	5.86	5.80	5.77	5.75
5	6.61	5.79	5.41	5.19	5.05	4.95	4.88	4.82	4.77	4.74	4.68	4.62	4.56	4.53	4.50
6	5.99	5.14	4.76	4.53	4.39	4.28	4.21	4.15	4.10	4.06	4.00	3.94	3.87	3.84	3.81
7	5.59	4.74	4.35	4.12	3.97	3.87	3.79	3.73	3.68	3.64	3.57	3.51	3.44	3.41	3.38
8	5.32	4.46	4.07	3.84	3.69	3.58	3.50	3.44	3.39	3.35	3.28	3.22	3.15	3.12	3.08
9	5.12	4.26	3.86	3.63	3.48	3.37	3.29	3.23	3.18	3.14	3.07	3.01	2.94	2.90	2.86
10	4.96	4.10	3.71	3.48	3.33	3.22	3.14	3.07	3.02	2.98	2.91	2.85	2.77	2.74	2.70
11	4.84	3.98	3.59	3.36	3.20	3.09	3.01	2.95	2.90	2.85	2.79	2.72	2.65	2.61	2.57
12	4.75	3.89	3.49	3.26	3.11	3.00	2.91	2.85	2.80	2.75	2.69	2.62	2.54	2.51	2.47
13	4.67	3.81	3.41	3.18	3.03	2.92	2.83	2.77	2.71	2.67	2.60	2.53	2.46	2.42	2.38
14	4.60	3.74	3.34	3.11	2.96	2.85	2.76	2.70	2.65	2.60	2.53	2.46	2.39	2.35	2.31
15	4.54	3.68	3.29	3.06	2.90	2.79	2.71	2.64	2.59	2.54	2.48	2.40	2.33	2.29	2.25
16	4.49	3.63	3.24	3.01	2.85	2.74	2.66	2.59	2.54	2.49	2.42	2.35	2.28	2.24	2.19
17	4.45	3.59	3.20	2.96	2.81	2.70	2.61	2.55	2.49	2.45	2.38	2.31	2.23	2.19	2.15
18	4.41	3.55	3.16	2.93	2.77	2.66	2.58	2.51	2.46	2.41	2.34	2.27	2.19	2.15	2.11
19	4.38	3.52	3.13	2.90	2.74	2.63	2.54	2.48	2.42	2.38	2.31	2.23	2.16	2.11	2.07

continued

Appendix A3A continued:

Degrees of freedom of lesser mean square (denominator)	Value of F for 5% probability and with the greater mean square having degrees of freedom:														
	1	2	3	4	5	6	7	8	9	10	12	15	20	24	30
20	4.35	3.49	3.10	2.87	2.71	2.60	2.51	2.45	2.39	2.35	2.28	2.20	2.12	2.08	2.04
21	4.32	3.47	3.07	2.84	2.68	2.57	2.49	2.42	2.37	2.32	2.25	2.18	2.10	2.05	2.01
22	4.30	3.44	3.05	2.82	2.66	2.55	2.46	2.40	2.34	2.30	2.23	2.15	2.07	2.03	1.98
23	4.28	3.42	3.03	2.80	2.64	2.53	2.44	2.37	2.32	2.27	2.20	2.13	2.05	2.01	1.96
24	4.26	3.40	3.01	2.78	2.62	2.51	2.42	2.36	2.30	2.25	2.18	2.11	2.03	1.98	1.94
25	4.24	3.39	2.99	2.76	2.60	2.49	2.40	2.34	2.28	2.24	2.16	2.09	2.01	1.96	1.92
26	4.23	3.37	2.98	2.74	2.59	2.47	2.39	2.32	2.27	2.22	2.15	2.07	1.99	1.95	1.90
27	4.21	3.35	2.96	2.73	2.57	2.46	2.37	2.31	2.25	2.20	2.13	2.06	1.97	1.93	1.88
28	4.20	3.34	2.95	2.71	2.56	2.45	2.36	2.29	2.24	2.19	2.12	2.04	1.96	1.91	1.87
29	4.18	3.33	2.93	2.70	2.55	2.43	2.35	2.28	2.22	2.18	2.10	2.03	1.94	1.90	1.85
30	4.17	3.32	2.92	2.69	2.53	2.42	2.33	2.27	2.21	2.16	2.09	2.01	1.93	1.89	1.84
40	4.08	3.23	2.84	2.61	2.45	2.34	2.25	2.18	2.12	2.08	2.00	1.92	1.84	1.79	1.74
60	4.00	3.15	2.76	2.53	2.37	2.25	2.17	2.10	2.04	1.99	1.92	1.84	1.75	1.70	1.65
120	3.92	3.07	2.68	2.45	2.29	2.17	2.09	2.02	1.96	1.91	1.83	1.75	1.66	1.61	1.55
∞	3.84	3.00	2.60	2.37	2.21	2.10	2.01	1.94	1.88	1.83	1.75	1.67	1.57	1.52	1.46

Appendix A3B. The Variance Ratio (F): 1% Significance Level

Degrees of freedom of lesser mean square (denominator)	Value of F for 1% probability and with the greater mean square having degrees of freedom:														
	1	2	3	4	5	6	7	8	9	10	12	15	20	24	30
1	4052	4999.5	5403	5625	5764	5859	5928	5981	6022	6056	6106	6157	6209	6235	6261
2	98.50	99.00	99.17	99.25	99.30	99.33	99.36	99.37	99.39	99.40	99.42	99.43	99.45	99.46	99.47
3	34.12	30.82	29.46	28.71	28.24	27.91	27.67	27.49	27.35	27.23	27.05	26.87	26.69	26.60	26.50
4	21.20	18.00	16.69	15.98	15.52	15.21	14.98	14.80	14.66	14.55	14.37	14.20	14.02	13.93	13.84
5	16.26	13.27	12.06	11.39	10.97	10.67	10.46	10.29	10.16	10.05	9.89	9.72	9.55	9.47	9.38
6	13.75	10.92	9.78	9.15	8.75	8.47	8.26	8.10	7.98	7.87	7.72	7.56	7.40	7.31	7.23
7	12.25	9.55	8.45	7.85	7.46	7.19	6.99	6.84	6.72	6.62	6.47	6.31	6.16	6.07	5.99
8	11.26	8.65	7.59	7.01	6.63	6.37	6.18	6.03	5.91	5.81	5.67	5.52	5.36	5.28	5.20
9	10.56	8.02	6.99	6.42	6.06	5.80	5.61	5.47	5.35	5.26	5.11	4.96	4.81	4.73	4.65
10	10.04	7.56	6.55	5.99	5.64	5.39	5.20	5.06	4.94	4.85	4.71	4.56	4.41	4.33	4.25
11	9.65	7.21	6.22	5.67	5.32	5.07	4.89	4.74	4.63	4.54	4.40	4.25	4.10	4.02	3.94
12	9.33	6.93	5.95	5.41	5.06	4.82	4.64	4.50	4.39	4.30	4.16	4.01	3.86	3.78	3.70
13	9.07	6.70	5.74	5.21	4.86	4.62	4.44	4.30	4.19	4.10	3.96	3.82	3.66	3.59	3.51
14	8.86	6.51	5.56	5.04	4.69	4.46	4.28	4.14	4.03	3.94	3.80	3.66	3.51	3.43	3.35
15	8.68	6.36	5.42	4.89	4.56	4.32	4.14	4.00	3.89	3.80	3.67	3.52	3.37	3.29	3.21
16	8.53	6.23	5.29	4.77	4.44	4.20	4.03	3.89	3.78	3.69	3.55	3.41	3.26	3.18	3.10
17	8.40	6.11	5.18	4.67	4.34	4.10	3.93	3.79	3.68	3.59	3.46	3.31	3.16	3.08	3.00
18	8.29	6.01	5.09	4.58	4.25	4.01	3.84	3.71	3.60	3.51	3.37	3.23	3.08	3.00	2.92
19	8.18	5.93	5.01	4.50	4.17	3.94	3.77	3.63	3.52	3.43	3.30	3.15	3.00	2.92	2.84

continued

Appendix A3B *continued*:

Degrees of freedom of lesser mean square (denominator)	Value of F for 1% probability and with the greater mean square having degrees of freedom:														
	1	2	3	4	5	6	7	8	9	10	12	15	20	24	30
20	8.10	5.85	4.94	4.43	4.10	3.87	3.70	3.56	3.46	3.37	3.23	3.09	2.94	2.86	2.78
21	8.02	5.78	4.87	4.37	4.04	3.81	3.64	3.51	3.40	3.31	3.17	3.03	2.88	2.80	2.72
22	7.95	5.72	4.82	4.31	3.99	3.76	3.59	3.45	3.35	3.26	3.12	2.98	2.83	2.75	2.67
23	7.88	5.66	4.76	4.26	3.94	3.71	3.54	3.41	3.30	3.21	3.07	2.93	2.78	2.70	2.62
24	7.82	5.61	4.72	4.22	3.90	3.67	3.50	3.36	3.26	3.17	3.03	2.89	2.74	2.66	2.58
25	7.77	5.57	4.68	4.18	3.85	3.63	3.46	3.32	3.22	3.13	2.99	2.85	2.70	2.62	2.54
26	7.72	5.53	4.64	4.14	3.82	3.59	3.42	3.29	3.18	3.09	2.96	2.81	2.66	2.58	2.50
27	7.68	5.49	4.60	4.11	3.78	3.56	3.39	3.26	3.15	3.06	2.93	2.78	2.63	2.55	2.47
28	7.64	5.45	4.57	4.07	3.75	3.53	3.36	3.23	3.12	3.03	2.90	2.75	2.60	2.52	2.44
29	7.60	5.42	4.54	4.04	3.73	3.50	3.33	3.20	3.09	3.00	2.87	2.73	2.57	2.49	2.41
30	7.56	5.39	4.51	4.02	3.70	3.47	3.30	3.17	3.07	2.98	2.84	2.70	2.55	2.47	2.39
40	7.31	5.18	4.31	3.83	3.51	3.29	3.12	2.99	2.89	2.80	2.66	2.52	2.37	2.29	2.20
60	7.08	4.98	4.13	3.65	3.34	3.12	2.95	2.82	2.72	2.63	2.50	2.35	2.20	2.12	2.03
120	6.85	4.79	3.95	3.48	3.17	2.96	2.79	2.66	2.56	2.47	2.34	2.19	2.03	1.95	1.86
∞	6.63	4.61	3.78	3.32	3.02	2.80	2.64	2.51	2.41	2.32	2.18	2.04	1.88	1.79	1.70

Appendix A4. The χ^2 Statistic

Degrees of freedom	Value of χ^2 for probability (%)			
	10	5	1	0.1
1	2.706	3.841	6.635	10.827
2	4.605	5.991	9.210	13.815
3	6.251	7.815	11.345	16.266
4	7.779	9.488	13.277	18.467
5	9.236	11.070	15.086	20.515
6	10.645	12.592	16.812	22.457
7	12.017	14.067	18.475	24.322
8	13.362	15.507	20.090	26.125
9	14.684	16.919	21.666	27.877
10	15.987	18.307	23.209	29.588
11	17.275	19.675	24.725	31.264
12	18.549	21.026	26.217	32.909
13	19.812	22.362	27.688	34.528
14	21.064	23.685	29.141	36.123
15	22.307	24.996	30.578	37.697
16	23.542	26.296	32.000	39.252
17	24.769	27.587	33.409	40.790
18	25.989	28.869	34.805	42.312
19	27.204	30.144	36.191	43.820
20	28.412	31.410	37.566	45.315
21	29.615	32.671	38.932	46.797
22	30.813	33.924	40.289	48.268
23	32.007	35.172	41.638	49.728
24	33.196	36.415	42.980	51.179
25	34.382	37.652	44.314	52.620
26	35.563	38.885	45.642	54.052
27	36.741	40.113	46.963	55.476
28	37.916	41.337	48.278	56.893
29	39.087	42.557	49.588	58.302
30	40.256	43.773	50.892	59.703
32	42.585	46.194	53.486	62.487
34	44.903	48.602	56.061	65.247
36	47.212	50.999	58.619	67.985
38	49.513	53.384	61.162	70.703
40	51.805	55.759	63.691	73.402
42	54.090	58.124	66.206	76.084
44	56.369	60.481	68.710	78.750
46	58.641	62.830	71.201	81.400
48	60.907	65.171	73.683	84.037
50	63.167	67.505	76.154	86.661

Appendix A5. Rankits

Note that rankit values are symmetrical about the median rank. To save space, negative values are not tabulated for samples where $N > 10$.

Rank order	Size of sample = N								
	2	3	4	5	6	7	8	9	10
1	0.564	0.864	1.029	1.163	1.267	1.352	1.424	1.485	1.539
2	-0.564	0.000	0.297	0.495	0.642	0.757	0.852	0.932	1.001
3		-0.864	-0.297	0.000	0.202	0.353	0.473	0.572	0.656
4			-1.029	-0.495	-0.202	0.000	0.153	0.275	0.376
5				-1.163	-0.642	-0.353	-0.153	0.000	0.123
6					-1.267	-0.757	-0.473	-0.275	-0.123
7						-1.352	-0.852	-0.572	-0.376
8							-1.424	-0.932	-0.656
9								-1.485	-1.001
10									-1.539

Rank order	11	12	13	14	15	16	17	18	19	20
1	1.586	1.629	1.668	1.703	1.736	1.766	1.794	1.820	1.844	1.867
2	1.062	1.116	1.164	1.208	1.248	1.285	1.319	1.350	1.380	1.408
3	0.729	0.793	0.850	0.901	0.948	0.990	1.029	1.066	1.099	1.131
4	0.462	0.537	0.603	0.662	0.715	0.763	0.807	0.848	0.886	0.921
5	0.225	0.312	0.388	0.456	0.516	0.570	0.619	0.665	0.707	0.745
6	0.000	0.103	0.191	0.267	0.335	0.396	0.451	0.502	0.548	0.590
7			0.000	0.088	0.165	0.234	0.295	0.351	0.402	0.448
8					0.000	0.077	0.146	0.208	0.264	0.315
9							0.000	0.069	0.131	0.187
10									0.000	0.062

continued

Appendix A5 *continued:*

	Size of sample = N									
Rank order	21	22	23	24	25	26	27	28	29	30
1	1.889	1.910	1.929	1.948	1.965	1.982	1.998	2.014	2.029	2.043
2	1.434	1.458	1.481	1.503	1.524	1.544	1.563	1.581	1.599	1.616
3	1.160	1.188	1.214	1.239	1.263	1.285	1.306	1.327	1.346	1.365
4	0.954	0.985	1.014	1.041	1.067	1.091	1.115	1.137	1.158	1.179
5	0.782	0.815	0.847	0.877	0.905	0.932	0.957	0.981	1.004	1.026
6	0.630	0.667	0.701	0.734	0.764	0.793	0.820	0.846	0.871	0.894
7	0.491	0.532	0.569	0.604	0.637	0.668	0.697	0.725	0.752	0.777
8	0.362	0.406	0.446	0.484	0.519	0.553	0.584	0.614	0.642	0.669
9	0.238	0.286	0.330	0.370	0.409	0.444	0.478	0.510	0.540	0.568
10	0.118	0.170	0.218	0.262	0.303	0.341	0.377	0.411	0.443	0.473
11	0.000	0.056	0.108	0.156	0.200	0.241	0.280	0.316	0.350	0.382
12			0.000	0.052	0.100	0.144	0.185	0.224	0.260	0.294
13					0.000	0.048	0.092	0.134	0.172	0.209
14							0.000	0.044	0.086	0.125
15									0.000	0.041

continued

Appendix A5 *continued*:

Rank order	\multicolumn{10}{c}{Size of sample $= N$}									
	31	32	33	34	35	36	37	38	39	40
1	2.056	2.070	2.082	2.095	2.107	2.118	2.129	2.140	2.151	2.161
2	1.632	1.647	1.662	1.676	1.690	1.704	1.717	1.729	1.741	1.753
3	1.383	1.400	1.416	1.432	1.448	1.462	1.477	1.491	1.504	1.517
4	1.198	1.217	1.235	1.252	1.269	1.285	1.300	1.315	1.330	1.344
5	1.047	1.067	1.087	1.105	1.123	1.140	1.157	1.173	1.188	1.203
6	0.917	0.938	0.959	0.979	0.998	1.016	1.034	1.051	1.067	1.083
7	0.801	0.824	0.846	0.867	0.887	0.906	0.925	0.943	0.960	0.977
8	0.694	0.719	0.742	0.764	0.786	0.806	0.826	0.845	0.863	0.881
9	0.595	0.621	0.646	0.670	0.692	0.714	0.735	0.755	0.774	0.793
10	0.502	0.529	0.556	0.580	0.604	0.627	0.649	0.670	0.690	0.710
11	0.413	0.442	0.469	0.496	0.521	0.545	0.568	0.590	0.611	0.632
12	0.327	0.358	0.387	0.414	0.441	0.466	0.490	0.514	0.536	0.557
13	0.243	0.276	0.307	0.336	0.364	0.390	0.416	0.440	0.463	0.486
14	0.161	0.196	0.228	0.259	0.289	0.317	0.343	0.369	0.393	0.417
15	0.080	0.117	0.151	0.184	0.215	0.245	0.273	0.300	0.325	0.350
16	0.000	0.039	0.076	0.110	0.143	0.174	0.203	0.232	0.258	0.284
17			0.000	0.037	0.071	0.104	0.135	0.165	0.193	0.220
18					0.000	0.035	0.067	0.099	0.128	0.156
19							0.000	0.033	0.064	0.094
20									0.000	0.031

From Bliss, C. I. (1967), *Statistics in Biology*, Vol. 1, p. 498–9. *Reproduced by permission of McGraw-Hill Book Company Ltd, New York.*

Appendix A6. Percentages and Probits

Percentage units	Probit corresponding to percentage decimals									
	0.0	0.1	0.2	0.3	0.4	0.5	0.6	0.7	0.8	0.9
0	...	1.9098	2.1218	2.2522	2.3479	2.4242	2.4879	2.5427	2.5911	2.6344
1	2.6737	2.7096	2.7429	2.7738	2.8027	2.8299	2.8556	2.8799	2.9031	2.9251
2	2.9463	2.9665	2.9859	3.0046	3.0226	3.0400	3.0569	3.0732	3.0890	3.1043
3	3.1192	3.1337	3.1478	3.1616	3.1750	3.1881	3.2009	3.2134	3.2256	3.2376
4	3.2493	3.2608	3.2721	3.2831	3.2940	3.3046	3.3151	3.3253	3.3354	3.3454
5	3.3551	3.3648	3.3742	3.3836	3.3928	3.4018	3.4107	3.4195	3.4282	3.4368
6	3.4452	3.4536	3.4618	3.4699	3.4780	3.4859	3.4937	3.5015	3.5091	3.5167
7	3.5242	3.5316	3.5389	3.5462	3.5534	3.5605	3.5675	3.5745	3.5813	3.5882
8	3.5949	3.6016	3.6083	3.6148	3.6213	3.6278	3.6342	3.6405	3.6468	3.6531
9	3.6592	3.6654	3.6715	3.6775	3.6835	3.6894	3.6953	3.7012	3.7070	3.7127
10	3.7184	3.7241	3.7298	3.7354	3.7409	3.7464	3.7519	3.7574	3.7628	3.7681
11	3.7735	3.7788	3.7840	3.7893	3.7945	3.7996	3.8048	3.8099	3.8150	3.8200
12	3.8250	3.8300	3.8350	3.8399	3.8448	3.8497	3.8545	3.8593	3.8641	3.8689
13	3.8736	3.8783	3.8830	3.8877	3.8923	3.8969	3.9015	3.9061	3.9107	3.9152
14	3.9197	3.9242	3.9286	3.9331	3.9375	3.9419	3.9463	3.9506	3.9550	3.9593
15	3.9636	3.9678	3.9721	3.9763	3.9806	3.9848	3.9890	3.9931	3.9973	4.0014
16	4.0055	4.0096	4.0137	4.0178	4.0218	4.0259	4.0299	4.0339	4.0379	4.0419
17	4.0458	4.0498	4.0537	4.0576	4.0615	4.0654	4.0693	4.0731	4.0770	4.0808
18	4.0846	4.0884	4.0922	4.0960	4.0998	4.1035	4.1073	4.1110	4.1147	4.1184
19	4.1221	4.1258	4.1295	4.1331	4.1367	4.1404	4.1440	4.1476	4.1512	4.1548
20	4.1584	4.1619	4.1655	4.1690	4.1726	4.1761	4.1796	4.1831	4.1866	4.1901
21	4.1936	4.1970	4.2005	4.2039	4.2074	4.2108	4.2142	4.2176	4.2210	4.2244
22	4.2278	4.2312	4.2345	4.2379	4.2412	4.2446	4.2479	4.2512	4.2546	4.2579
23	4.2612	4.2644	4.2677	4.2710	4.2743	4.2775	4.2808	4.2840	4.2872	4.2905
24	4.2937	4.2969	4.3001	4.3033	4.3065	4.3097	4.3129	4.3160	4.3192	4.3324

continued

Appendix A.6 continued:

Percentage units	Probit corresponding to percentage decimals									
	0.0	0.1	0.2	0.3	0.4	0.5	0.6	0.7	0.8	0.9
25	4.3255	4.3287	4.3318	4.3349	4.3380	4.3412	4.3443	4.3474	4.3505	4.3536
26	4.3567	4.3597	4.3628	4.3659	4.3689	4.3720	4.3750	4.3781	4.3811	4.3842
27	4.3872	4.3902	4.3932	4.3962	4.3992	4.4022	4.4052	4.4082	4.4112	4.4142
28	4.4172	4.4201	4.4231	4.4260	4.4290	4.4319	4.4349	4.4378	4.4408	4.4437
29	4.4466	4.4495	4.4524	4.4554	4.4583	4.4612	4.4641	4.4670	4.4698	4.4727
30	4.4756	4.4785	4.4813	4.4842	4.4871	4.4899	4.4928	4.4956	4.4985	4.5013
31	4.5041	4.5070	4.5098	4.5126	4.5155	4.5183	4.5211	4.5239	4.5267	4.5295
32	4.5323	4.5351	4.5379	4.5407	4.5435	4.5462	4.5490	4.5518	4.5546	4.5573
33	4.5601	4.5628	4.5656	4.5684	4.5711	4.5739	4.5766	4.5793	4.5821	4.5848
34	4.5875	4.5903	4.5930	4.5957	4.5984	4.6011	4.6039	4.6066	4.6093	4.6120
35	4.6147	4.6174	4.6201	4.6228	4.6255	4.6281	4.6308	4.6335	4.6362	4.6389
36	4.6415	4.6442	4.6469	4.6495	4.6522	4.6549	4.6575	4.6602	4.6628	4.6655
37	4.6681	4.6708	4.6734	4.6761	4.6787	4.6814	4.6840	4.6866	4.6893	4.6919
38	4.6945	4.6971	4.6998	4.7024	4.7050	4.7076	4.7102	4.7129	4.7155	4.7181
39	4.7207	4.7233	4.7259	4.7285	4.7311	4.7337	4.7363	4.7389	4.7415	4.7441
40	4.7467	4.7492	4.7518	4.7544	4.7570	4.7596	4.7622	4.7647	4.7673	4.7699
41	4.7725	4.7750	4.7776	4.7802	4.7827	4.7853	4.7879	4.7904	4.7930	4.7955
42	4.7981	4.8007	4.8032	4.8058	4.8083	4.8109	4.8134	4.8160	4.8185	4.8211
43	4.8236	4.8262	4.8287	4.8313	4.8338	4.8363	4.8389	4.8414	4.8440	4.8465
44	4.8490	4.8516	4.8541	4.8566	4.8592	4.8617	4.8642	4.8668	4.8693	4.8718
45	4.8743	4.8769	4.8794	4.8819	4.8844	4.8870	4.8895	4.8920	4.8945	4.8970
46	4.8996	4.9021	4.9046	4.9071	4.9096	4.9122	4.9147	4.9172	4.9197	4.9222
47	4.9247	4.9272	4.9298	4.9323	4.9348	4.9373	4.9398	4.9423	4.9448	4.9473
48	4.9498	4.9524	4.9549	4.9574	5.9599	4.9624	4.9649	4.9674	4.9699	4.9724
49	4.9749	4.9774	4.9799	4.9825	4.9850	4.9875	4.9900	4.9925	4.9950	4.9975

continued

Appendix A.6 *continued:*

| Percentage units | Probit corresponding to percentage decimals | | | | | | | | | |
	0.0	0.1	0.2	0.3	0.4	0.5	0.6	0.7	0.8	0.9
50	5.0000	5.0025	5.0050	5.0075	5.0100	5.0125	5.0150	5.0175	5.0201	5.0226
51	5.0251	5.0276	5.0301	5.0326	5.0351	5.0376	5.0401	5.0426	5.0451	5.0476
52	5.0502	5.0527	5.0552	5.0577	5.0602	5.0627	5.0652	5.0677	5.0702	5.0728
53	5.0753	5.0778	5.0803	5.0828	5.0853	5.0878	5.0904	5.0929	5.0954	5.0979
54	5.1004	5.1030	5.1055	5.1080	5.1105	5.1130	5.1156	5.1181	5.1206	5.1231
55	5.1257	5.1282	5.1307	5.1332	5.1358	5.1383	5.1408	5.1434	5.1459	5.1484
56	5.1510	5.1535	5.1560	5.1586	5.1611	5.1637	5.1662	5.1687	5.1713	5.1738
57	5.1764	5.1789	5.1815	5.1840	5.1866	5.1891	5.1917	5.1942	5.1968	5.1993
58	5.2019	5.2045	5.2070	5.2096	5.2121	5.2147	5.2173	5.2198	5.2224	5.2250
59	5.2275	5.2301	5.2327	5.2353	5.2378	5.2404	5.2430	5.2456	5.2482	5.2508
60	5.2533	5.2559	5.2585	5.2611	5.2637	5.2663	5.2689	5.2715	5.2741	5.2767
61	5.2793	5.2819	5.2845	5.2871	5.2898	5.2924	5.2950	5.2976	5.3002	5.3029
62	5.3055	5.3081	5.3107	5.3134	5.3160	5.3186	5.3213	5.3239	5.3266	5.3292
63	5.3319	5.3345	5.3372	5.3398	5.3425	5.3451	5.3478	5.3505	5.3531	5.3558
64	5.3585	5.3611	5.3638	5.3665	5.3692	5.3719	5.3745	5.3772	5.3799	5.3826
65	5.3853	5.3880	5.3907	5.3934	5.3961	5.3989	5.4016	5.4043	5.4070	5.4097
66	5.4125	5.4152	5.4179	5.4207	5.4234	5.4261	5.4289	5.4316	5.4344	5.4372
67	5.4399	5.4427	5.4454	5.4482	5.4510	5.4538	5.4565	5.4593	5.4621	5.4649
68	5.4677	5.4705	5.4733	5.4761	5.4789	5.4817	5.4845	5.4874	5.4902	5.4930
69	5.4959	5.4987	5.5015	5.5044	5.5072	5.5101	5.5129	5.5158	5.5187	5.5215
70	5.5244	5.5273	5.5302	5.5330	5.5359	5.5388	5.5417	5.5446	5.5476	5.5505
71	5.5534	5.5563	5.5592	5.5622	5.5651	5.5681	5.5710	5.5740	5.5769	5.5799
72	5.5828	5.5858	5.5888	5.5918	5.5948	5.5978	5.6008	5.6038	5.6068	5.6098
73	5.6128	5.6158	5.6189	5.6219	5.6250	5.6280	5.6311	5.6341	5.6372	5.6403
74	5.6433	5.6464	5.6495	5.6526	5.6557	5.6588	5.6620	5.6651	5.6682	5.6713

continued

Probit corresponding to percentage decimals

Percentage units	0.0	0.1	0.2	0.3	0.4	0.5	0.6	0.7	0.8	0.9
75	5.6745	5.6776	5.6808	5.6840	5.6871	5.6903	5.6935	5.6967	5.6999	5.7031
76	5.7063	5.7095	5.7128	5.7160	5.7192	5.7225	5.7257	5.7290	5.7323	5.7356
77	5.7388	5.7421	5.7454	5.7488	5.7521	5.7554	5.7588	5.7621	5.7655	5.7688
78	5.7722	5.7756	5.7790	5.7824	5.7858	5.7892	5.7926	5.7961	5.7995	5.8030
79	5.8064	5.8099	5.8134	5.8169	5.8204	5.8239	5.8274	5.8310	5.8345	5.8381
80	5.8416	5.8452	5.8488	5.8524	5.8560	5.8596	5.8633	5.8669	5.8705	5.8742
81	5.8779	5.8816	5.8853	5.8890	5.8927	5.8965	5.9002	5.9040	5.9078	5.9116
82	5.9154	5.9192	5.9230	5.9269	5.9307	5.9346	5.9385	5.9424	5.9463	5.9502
83	5.9542	5.9581	5.9621	5.9661	5.9701	5.9741	5.9782	5.9822	5.9863	5.9904
84	5.9945	5.9986	6.0027	6.0069	6.0110	6.0152	6.0194	6.0237	6.0279	6.0322
85	6.0364	6.0407	6.0450	6.0494	6.0537	6.0581	6.0625	6.0669	6.0714	6.0758
86	6.0803	6.0848	6.0893	6.0939	6.0985	6.1031	6.1077	6.1123	6.1170	6.1217
87	6.1264	6.1311	6.1359	6.1407	6.1455	6.1503	6.1552	6.1601	6.1650	6.1700
88	6.1750	6.1800	6.1850	6.1901	6.1952	6.2004	6.2055	6.2107	6.2160	6.2212
89	6.2265	6.2319	6.2372	6.2426	6.2481	6.2536	6.2591	6.2646	6.2702	6.2759
90	6.2816	6.2873	6.2930	6.2988	6.3047	6.3106	6.3165	6.3225	6.3285	6.3346
91	6.3408	6.3469	6.3532	6.3595	6.3658	6.3722	6.3787	6.3852	6.3917	6.3984
92	6.4051	6.4118	6.4187	6.4255	6.4325	6.4395	6.4466	6.4538	6.4611	6.4684
93	6.4758	6.4833	6.4909	6.4985	6.5063	6.5141	6.5220	6.5301	6.5382	6.5464
94	6.5548	6.5632	6.5718	6.5805	6.5893	6.5982	6.6072	6.6164	6.6258	6.6352
95	6.6449	6.6546	6.6646	6.6747	6.6849	6.6954	6.7060	6.7169	6.7279	6.7392
96	6.7507	6.7624	6.7744	6.7866	6.7991	6.8119	6.8250	6.8384	6.8522	6.8663
97	6.8808	6.8957	6.9110	6.9268	6.9431	6.9600	6.9774	6.9954	7.0141	7.0335
98.0	7.0537	7.0558	7.0579	7.0600	7.0621	7.0642	7.0663	7.0684	7.0706	7.0727
98.1	7.0749	7.0770	7.0792	7.0814	7.0836	7.0858	7.0880	7.0902	7.0924	7.0947
98.2	7.0969	7.0992	7.1015	7.1038	7.1061	7.1084	7.1107	7.1130	7.1154	7.1177
98.3	7.1201	7.1224	7.1248	7.1272	7.1297	7.1321	7.1345	7.1370	7.1394	7.1419
98.4	7.1444	7.1469	7.1494	7.1520	7.1545	7.1571	7.1596	7.1622	7.1648	7.1675

continued

Appendix A.6 continued:

Percentage units	Probit corresponding to percentage decimals									
	0.0	0.1	0.2	0.3	0.4	0.5	0.6	0.7	0.8	0.9
98.5	7.1701	7.1727	7.1754	7.1781	7.1808	7.1835	7.1862	7.1890	7.1917	7.1945
98.6	7.1973	7.2001	7.2029	7.2058	7.2086	7.2115	7.2144	7.2173	7.2203	7.2232
98.7	7.2262	7.2292	7.2322	7.2353	7.2383	7.2414	7.2445	7.2476	7.2508	7.2539
98.8	7.2571	7.2603	7.2636	7.2668	7.2701	7.2734	7.2768	7.2801	7.2835	7.2869
98.9	7.2904	7.2938	7.2973	7.3009	7.3044	7.3080	7.3116	7.3152	7.3189	7.3226
99.0	7.3263	7.3301	7.3339	7.3378	7.3416	7.3455	7.3495	7.3535	7.3575	7.3615
99.1	7.3656	7.3698	7.3739	7.3781	7.3824	7.3867	7.3911	7.3954	7.3999	7.4044
99.2	7.4089	7.4135	7.4181	7.4228	7.4276	7.4324	7.4372	7.4422	7.4471	7.4522
99.3	7.4573	7.4624	7.4677	7.4730	7.4783	7.4838	7.4893	7.4949	7.5006	7.5063
99.4	7.5121	7.5181	7.5241	7.5302	7.5364	7.5427	7.5491	7.5556	7.5622	7.5690
99.5	7.5758	7.5828	7.5899	7.5972	7.6045	7.6121	7.6197	7.6276	7.6356	7.6437
99.6	7.6521	7.6606	7.6693	7.6783	7.6874	7.6968	7.7065	7.7164	7.7266	7.7370
99.7	7.7478	7.7589	7.7703	7.7822	7.7944	7.8070	7.8202	7.8338	7.8480	7.8627
99.8	7.8782	7.8943	7.9112	7.9290	7.9478	7.9677	7.9889	8.0115	8.0357	8.0618
99.9	8.0902	8.1214	8.1559	8.1947	8.2389	8.2905	8.3528	8.4316	8.5401	8.7190

Appendix A7. Weighting Coefficients for Expected Probits

Expected probit			Weighting coefficient
2.0	or	8.0	0.015
2.1		7.9	0.019
2.2		7.8	0.025
2.3		7.7	0.031
2.4		7.6	0.040
2.5	or	7.5	0.050
2.6		7.4	0.062
2.7		7.3	0.076
2.8		7.2	0.092
2.9		7.1	0.110
3.0	or	7.0	0.131
3.1		6.9	0.154
3.2		6.8	0.180
3.3		6.7	0.208
3.4		6.6	0.237
3.5	or	6.5	0.269
3.6		6.4	0.302
3.7		6.3	0.336
3.8		6.2	0.370
3.9		6.1	0.405
4.0	or	6.0	0.439
4.1		5.9	0.471
4.2		5.8	0.503
4.3		5.7	0.532
4.4		5.6	0.558
4.5	or	5.5	0.581
4.6		5.4	0.601
4.7		5.3	0.616
4.8		5.2	0.627
4.9		5.1	0.634
	5.0		0.637

Appendix A8. Confidence Limits from a Single Count

	Confidence limits (%)	
Count	95	99
0	0 – 3.6889	0 – 5.2983
1	0.0253– 5.5716	0.0050– 7.4301
2	0.2422– 7.2247	0.1035– 9.2738
3	0.6187– 8.7673	0.3379–10.978
4	1.0899–10.242	0.6722–12.595
5	1.6235–11.669	1.0779–14.150
6	2.2019–13.060	1.5369–15.660
7	2.8144–14.423	2.0374–17.134
8	3.4539–15.764	2.5711–18.579
9	4.1154–17.085	3.1325–19.999
10	4.7954–18.391	3.7172–21.398
11	5.4913–19.683	4.3216–22.780
12	6.2008–20.962	4.9434–24.145
13	6.9223–22.231	5.5807–25.497
14	7.6542–23.490	6.2316–26.836
15	8.3957–24.741	6.8946–28.164
16	9.1459–25.983	7.5680–29.482
17	9.9037–27.219	8.2518–30.791
18	10.668 –28.448	8.9453–32.091
19	11.440 –29.671	9.6470–33.383
20	12.217 –30.889	10.355 –34.668
21	13.000 –32.101	11.072 –35.947
22	13.788 –33.309	11.795 –37.219
23	14.581 –34.512	12.525 –38.485
24	15.378 –35.711	13.260 –39.745
25	16.178 –36.905	14.000 –41.001
26	16.983 –38.097	14.745 –42.251
27	17.793 –39.284	15.495 –43.497
28	18.606 –40.468	16.244 –44.739
29	19.422 –41.649	17.003 –45.976
30	20.241 –42.827	17.767 –47.210
31	21.063 –44.002	18.535 –48.439
32	21.888 –45.175	19.306 –49.666
33	22.715 –46.345	20.080 –50.888
34	23.545 –47.512	20.858 –52.108
35	24.378 –48.677	21.638 –53.324
36	25.213 –49.840	22.422 –54.538
37	26.050 –51.000	23.208 –55.748
38	26.890 –52.158	23.997 –56.956
39	27.732 –53.315	24.789 –58.161
40	28.575 –54.469	25.583 –59.363
41	29.421 –55.622	26.381 –60.564
42	30.269 –56.772	27.181 –61.761
43	31.119 –57.921	27.983 –62.957
44	31.970 –59.068	28.788 –64.150
45	32.823 –60.214	29.596 –65.341
46	33.678 –61.358	30.406 –66.530
47	34.534 –62.501	31.218 –67.717
48	35.392 –63.642	32.032 –68.902
49	36.251 –64.781	32.848 –70.085

continued:

Appendix A8. *continued:*

	Confidence limits (%)	
Count	95	99
50	37.112– 65.919	33.666– 71.267
51	37.973– 67.056	34.485– 72.446
52	38.837– 68.192	35.306– 73.624
53	39.701– 69.326	36.129– 74.800
54	40.567– 70.459	36.953– 75.975
55	41.433– 71.591	37.779– 77.148
56	42.301– 72.721	38.605– 78.319
57	43.171– 73.851	39.434– 79.489
58	44.041– 74.979	40.263– 80.657
59	44.912– 76.106	41.094– 81.825
60	45.785– 77.232	41.926– 82.990
61	46.658– 78.357	42.759– 84.155
62	47.533– 79.482	43.594– 85.317
63	48.409– 80.605	44.430– 86.479
64	49.286– 81.727	45.267– 87.640
65	50.164– 82.848	46.106– 88.799
66	51.042– 83.969	46.946– 89.957
67	51.922– 85.088	47.787– 91.114
68	52.803– 86.207	48.630– 92.269
69	53.685– 87.324	49.475– 93.424
70	54.567– 88.441	50.320– 94.577
71	55.451– 89.557	51.167– 95.730
72	56.335– 90.673	52.015– 96.881
73	57.220– 91.787	52.865– 98.031
74	58.106– 92.901	53.716– 99.180
75	58.993– 94.014	54.567–100.33
76	59.880– 95.126	55.420–101.48
77	60.768– 96.237	56.275–102.63
78	61.657– 97.348	57.130–103.77
79	62.547– 98.458	57.986–104.92
80	63.437– 99.567	58.844–106.06
81	64.328–100.68	59.701–107.20
82	65.219–101.79	60.559–108.34
83	66.111–102.90	61.419–109.49
84	67.003–104.00	62.279–110.63
85	67.897–105.11	63.140–111.77
86	68.790–106.21	64.001–112.90
87	69.684–107.32	64.863–114.04
88	70.579–108.42	65.725–115.18
89	71.474–109.53	66.587–116.31
90	72.370–110.63	67.451–117.45
91	73.267–111.73	68.314–118.59
92	74.164–112.83	69.179–119.72
93	75.061–113.94	70.043–120.85
94	75.959–115.04	70.909–121.98
95	76.858–116.14	71.775–123.12
96	77.757–117.24	72.641–124.25
97	78.657–118.34	73.508–125.38
98	79.557–119.44	74.375–126.51
99	80.458–120.53	75.244–127.64
100	81.36 –121.66	75.90 –128.55

From *Documenta Geigy Scientific Tables, 7th edn*, p. 107, 108. *Reproduced with kind permission of Ciba-Geigy Ltd, Basle (Switzerland).*

Appendix A9. Most Probable Numbers from 3 Successive 10-fold Dilutions, 5 Tubes of Each

Numbers of tubes positive out of 5 at 3 successive dilutions			MPN per inoculum of the first dilution
0	1	0	0.18
1	0	0	0.20
1	1	0	0.40
2	0	0	0.45
2	0	1	0.68
2	1	0	0.68
2	2	0	0.93
3	0	0	0.78
3	0	1	1.1
3	1	0	1.1
3	2	0	1.4
4	0	0	1.3
4	0	1	1.7
4	1	0	1.7
4	1	1	2.1
4	2	0	2.2
4	2	1	2.6
4	3	0	2.7
5	0	0	2.3
5	0	1	3.1
5	1	0	3.3
5	1	1	4.6
5	2	0	4.9
5	2	1	7.0
5	2	2	9.5
5	3	0	7.9
5	3	1	11.0
5	3	2	14.0
5	4	0	13.0
5	4	1	17.0
5	4	2	22.0
5	4	3	28.0
5	5	0	24.0
5	5	1	35.0
5	5	2	54.0
5	5	3	92.0
5	5	4	160.0

Adapted from Taylor, J. (1962). *Reproduced by permission of the Society for Applied Bacteriology.*

Appendix A10. Most Probable Numbers from 3 Successive 10-fold Dilutions, 8 Tubes of Each

Numbers of tubes positive out of 8 at 3 successive dilutions			MPN per inoculum of the first dilution
8	8	7	208
8	8	6	139
8	8	5	98.2
8	8	4	70.2
8	8	3	51.0
8	8	2	38.5
8	8	1	30.1
8	8	0	24.0
8	7	8	59.6
8	7	7	50.8
8	7	6	43.3
8	7	5	36.9
8	7	4	31.4
8	7	3	26.7
8	7	2	22.6
8	7	1	19.1
8	7	0	15.9
8	6	6	28.4
8	6	5	25.0
8	6	4	21.8
8	6	3	18.9
8	6	2	16.3
8	6	1	13.8
8	6	0	11.5
8	5	6	21.3
8	5	5	18.9
8	5	4	16.6
8	5	3	14.4
8	5	2	12.3
8	5	1	10.30
8	5	0	8.42
8	4	5	14.8
8	4	4	13.0
8	4	3	11.1
8	4	2	9.40
8	4	1	7.74
8	4	0	6.22
8	3	5	11.8
8	3	4	10.2
8	3	3	8.67
8	3	2	7.18
8	3	1	5.82

(continued)

Appendix A10 *continued:*

Number of tubes positive out of 8 at 3 successive dilutions			MPN per inoculum of the first dilution
8	3	0	4.67
8	2	4	8.07
8	2	3	6.72
8	2	2	5.50
8	2	1	4.45
8	2	0	3.62
8	1	3	5.22
8	1	2	4.27
8	1	1	3.50
8	1	0	2.87
8	0	2	3.38
8	0	1	2.80
8	0	0	2.31
7	7	1	5.47
7	7	0	4.84
7	6	2	5.30
7	6	1	4.71
7	6	0	4.15
7	5	2	4.58
7	5	1	4.04
7	5	0	3.55
7	4	3	4.46
7	4	2	3.95
7	4	1	3.47
7	4	0	3.04
7	3	3	3.86
7	3	2	3.40
7	3	1	2.98
7	3	0	2.59
7	2	3	3.33
7	2	2	2.92
7	2	1	2.55
7	2	0	2.20
7	1	3	2.87
7	1	2	2.51
7	1	1	2.17
7	1	0	1.86
7	0	2	2.14
7	0	1	1.83
7	0	0	1.55
6	6	1	3.08
6	6	0	2.77
6	5	1	2.73

(continued)

Appendix A10 *continued:*

Numbers of tubes positive out of 8 at 3 successive dilutions			MPN per inoculum of the first dilution
6	5	0	2.44
6	4	2	2.69
6	4	1	2.41
6	4	0	2.14
6	3	2	2.38
6	3	1	2.11
6	3	0	1.86
6	2	2	2.09
6	2	1	1.84
6	2	0	1.60
6	1	2	1.82
6	1	1	1.58
6	1	0	1.35
6	0	2	1.56
6	0	1	1.34
6	0	0	1.13
5	5	1	2.07
5	5	0	1.85
5	4	1	1.84
5	4	0	1.63
5	3	2	1.82
5	3	1	1.61
5	3	0	1.41
5	2	2	1.60
5	2	1	1.40
5	2	0	1.21
5	1	2	1.39
5	1	1	1.20
5	1	0	1.01
5	0	2	1.19
5	0	1	1.01
5	0	0	0.83
4	4	0	1.28
4	3	1	1.27
4	3	0	1.10
4	2	1	1.09
4	2	0	0.93
4	1	2	1.08
4	1	1	0.92
4	1	0	0.76
4	0	2	0.91
4	0	1	0.75
4	0	0	0.60

(continued)

Appendix A10 *continued:*

Numbers of tubes positive out of 8 at 3 successive dilutions			MPN per inoculum of the first dilution
3	4	0	1.01
3	3	1	1.00
3	3	0	0.85
3	2	1	0.85
3	2	0	0.70
3	1	2	0.84
3	1	1	0.70
3	1	0	0.56
3	0	2	0.69
3	0	1	0.55
3	0	0	0.41
2	4	0	0.79
2	3	1	0.79
2	3	0	0.66
2	2	1	0.65
2	2	0	0.52
2	1	1	0.52
2	1	0	0.39
2	0	2	0.51
2	0	1	0.38
2	0	0	0.26
1	3	0	0.49
1	2	1	0.49
1	2	0	0.36
1	1	1	0.36
1	1	0	0.24
1	0	2	0.36
1	0	1	0.24
1	0	0	0.12
0	2	0	0.23
0	1	1	0.23
0	1	0	0.11
0	0	1	0.11

From Norman, R. L. and Kempe, L. L. (1960), *J. biochem. microbiol. technol. Engng.,* **2,** 157–163. *Reproduced by permission of John Wiley & Sons Inc.*

Appendix A11. Significance of Differences between Proportions

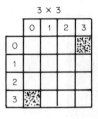

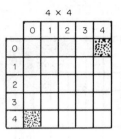

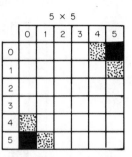

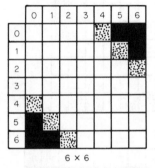

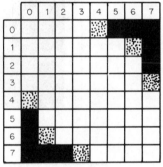

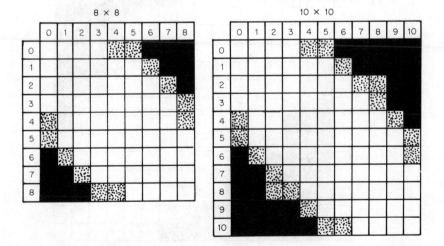

Appendix A.11A   Diagrams for determining the significance of differences between proportions from 3×3 to 10×10 comparisons. Black areas: $P \leqslant 1\%$; dotted areas: $5\% \geqslant P > 1\%$; plain areas: $P > 5\%$ (difference not significant)

Appendix A.11 *continued:*

12 × 12

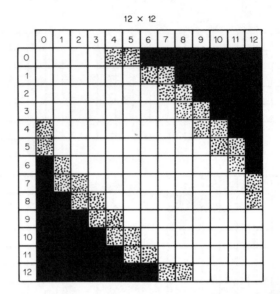

15 × 15

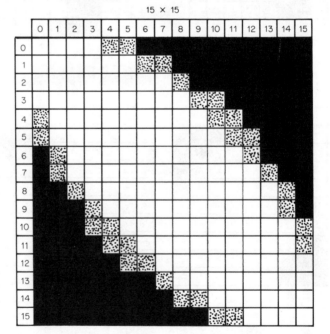

Appendix A.11B (continued): diagrams for 12 × 12 and 15 × 15 comparisons

Appendix A12. The U-Statistic for the Mann–Whitney U-Test

The table shows the critical values of U needed for groups of N_1 and N_2 observations to differ in their medians at the 5% level of significance.

N_1/N_2	1	2	3	4	5	6	7	8	9	10	11	12	13	14	15	16	17	18	19	20
1																				
2								0	0	0	0	1	1	1	1	1	2	2	2	2
3					0	1	1	2	2	3	3	4	4	5	5	6	6	7	7	8
4				0	1	2	3	4	4	5	6	7	8	9	10	11	11	12	13	13
5			0	1	2	3	5	6	7	8	9	11	12	13	14	15	17	18	19	20
6			1	2	3	5	6	8	10	11	13	14	16	17	19	21	22	24	25	27
7			1	3	5	6	8	10	12	14	16	18	20	22	24	26	28	30	32	34
8		0	2	4	6	8	10	13	15	17	19	22	24	26	29	31	34	36	38	41
9		0	2	4	7	10	12	15	17	20	23	26	28	31	34	37	39	42	45	48
10		0	3	5	8	11	14	17	20	23	26	29	33	36	39	42	45	48	52	55
11		0	3	6	9	13	16	19	23	26	30	33	37	40	44	47	51	55	58	62
12		1	4	7	11	14	18	22	26	29	33	37	41	45	49	53	57	61	65	69
13		1	4	8	12	16	20	24	28	33	37	41	45	50	54	59	63	67	72	76
14		1	5	9	13	17	22	26	31	36	40	45	50	55	59	64	67	74	78	83
15		1	5	10	14	19	24	29	34	39	44	49	54	59	64	70	75	80	85	90
16		1	6	11	15	21	26	31	37	42	47	53	59	64	70	75	81	86	92	98
17		2	6	11	17	22	28	34	39	45	51	57	63	67	75	81	87	93	99	105
18		2	7	12	18	24	30	36	42	48	55	61	67	74	80	86	93	99	106	112
19		2	7	13	19	25	32	38	45	52	58	65	72	78	85	92	99	106	113	119
20		2	8	13	20	27	34	41	48	55	62	69	76	83	90	98	105	112	119	127

From Campbell, R. C. (1974), *Statistics for Biologists*, 2nd edn, p. 347. *Reproduced by permission of Cambridge University Press.*

Appendix A13. The T-Statistic for the Wilcoxon Signed Rank Test

No. of paired samples (N)	Value of T for significance at (%)	
	5	1
6	0	
7	2	
8	4	0
9	6	2
10	8	3
11	11	5
12	14	7
13	17	10
14	21	13
15	25	16
16	30	20
17	35	23
18	40	28
19	46	32
20	52	38
21	59	43
22	66	49
23	73	55
24	81	61
25	89	68

From Siegel, S. (1956), *Nonparametric Statistics for the Behavioral Sciences*, p. 254. *Reproduced by permission of McGraw-Hill Book Company, New York.*

Appendix A14. The *R*-Statistic for the Dixon and Mood Sign Test

No. of informative pairs (N)	Value of R for significance at (%)	
	5	1
6	0	
7	0	
8	0	0
9	1	0
10	1	0
11	1	0
12	2	1
13	2	1
14	2	1
15	3	2
16	3	2
17	4	2
18	4	3
19	4	3
20	5	3
21	5	4
22	5	4
23	6	4
24	6	5
25	7	5

From Campbell, R. C. (1974), *Statistics for Biologists*, 2nd edn, p. 352. *Reproduced by permission of Cambridge University Press.*

Appendix A15. The Kolmogorov–Smirnov Statistic (K) (for the 2-sample test)

No. of observations in each group	Value of K for significance at level (%)	
	5	1
3	3	—
4	4	—
5	4	5
6	5	6
7	5	6
8	5	6
9	6	7
10	6	7
11	6	8
12	6	8
13	7	8
14	7	8
15	7	9
16	7	9
17	8	9
18	8	10
19	8	10
20	8	10
21	8	10
22	9	11
23	9	11
24	9	11
25	9	11
26	9	11
27	9	12
28	10	12
29	10	12
30	10	12
35	11	13
40	11	14

Abridged from Siegel, S. (1956), *Nonparametric Statistics for the Behavioral Sciences*, p. 278. *Reproduced by permission of McGraw-Hill Book Company.*

Appendix A16. The Friedman Statistic (S)

No. of randomized blocks (m)	Critical value of S for number of treatments (n) and approximate significance (%)								
	n = 3			n = 4			n = 5		
	5	1	0.1	5	1	0.1	5	1	0.1
2				20 4.2					
3	18 2.8[a]			37 3.3			64 4.5	76 0.78	86 0.09
4	26 4.2	32 0.46		52 3.6	64 0.69	74 0.09			
5	32 3.9	42 0.85	50 0.08	65 4.4	83 0.87	105 0.06			
6	42 2.9	54 0.81	72 0.01	76 4.3	100 1.00	128 0.09			
7	50 2.7	62 0.84	86 0.03						
8	50 4.7	72 0.99	98 0.09						
9	56 4.8	78 1.00	114 0.07						
10	62 4.6	96 0.75	126 0.08						

From Campbell, R. C. (1947), *Statistics for Biologists*, 2nd edn, p.353. *Reproduced by permission of Cambridge University Press.*
[a]This and similar entries are the exact levels of significance (%).

Appendix A17. Product–Moment Correlation Coefficient (r)

Degrees of freedom	Value of r for significance at (%)		
	5	1	0.1
1	0.99692	0.999877	0.9999988
2	0.95000	0.990000	0.99900
3	0.8793	0.95873	0.99116
4	0.8114	0.91720	0.97406
5	0.7545	0.8745	0.95074
6	0.7067	0.8343	0.92493
7	0.6664	0.7977	0.8982
8	0.6319	0.7646	0.8721
9	0.6021	0.7348	0.8471
10	0.5760	0.7079	0.8233
11	0.5529	0.6835	0.8010
12	0.5324	0.6614	0.7800
13	0.5139	0.6411	0.7603
14	0.4973	0.6226	0.7420
15	0.4821	0.6055	0.7246
16	0.4683	0.5897	0.7084
17	0.4555	0.5751	0.6932
18	0.4438	0.5614	0.6787
19	0.4329	0.5487	0.6652
20	0.4227	0.5368	0.6524
25	0.3809	0.4869	0.5974
30	0.3494	0.4487	0.5541
35	0.3246	0.4182	0.5189
40	0.3044	0.3932	0.4896
45	0.2875	0.3721	0.4648
50	0.2732	0.3541	0.4433
60	0.2500	0.3248	0.4078
70	0.2319	0.3017	0.3799
80	0.2172	0.2830	0.3568
90	0.2050	0.2673	0.3375
100	0.1946	0.2540	0.3211

Index